GEOTECHNICAL ENGINEERING FOR TRANSPORTATION INFRASTRUCTURE
LA GEOTECHNIQUE DANS LES INFRASTRUCTURES DE TRANSPORT
ADDITIONAL VOLUME
VOLUME ADDITIONNEL

COMPTES RENDUS DU DOUZIEME CONGRES EUROPEEN DE MECANIQUE DES SOLS
ET DE LA GEOTECHNIQUE AMSTERDAM/PAYS-BAS/7-10 JUIN 1999

La Géotechnique dans les Infrastructures de Transport

Théorie et Pratique, Projet et Conception, Construction et Entretien

Rédacteurs
F.B.J.Barends, J.Lindenberg, H.J.Luger, L.de Quelerij & A.Verruijt

VOLUME ADDITIONNEL

A.A.BALKEMA/ROTTERDAM/BROOKFIELD/2000

"Proof," Policy, & Practice

Understanding the Role of Evidence in Improving Education

Paul E. Lingenfelter

Foreword by Michael S. McPherson

"This book, whose author has been deeply involved in enacting and analyzing higher education policy in the United States for over 2 decades, makes an important contribution to our understanding of a key question: How do we know what policies actually work? The volume articulately describes the disconnect between academic researchers, who often pursue their own questions of interest, and the policy world, which is desperate for usable knowledge to help guide important decisions. Drawing on examples from education and other disciplines, it helps us understand why this chasm exists and how it best can be bridged."
—***Donald E. Heller***, *Dean, College of Education, Michigan State University*

"Lingenfelter takes on the longstanding and highly problematic relationship among research, policy, and practice. He unmasks modern-day shibboleths about how performance management and randomized field trials are the new answers. While highly respectful of practitioner wisdom and judgment, he marks out clear limits here too. He argues persuasively that those engaged in the work of education must become active agents of its continuous improvement, and sketches out how policymakers can foster an environment where such systematic gathering and use of evidence is more likely to happen. This is a very wise book!"—***Anthony S. Bryk***, *President, Carnegie Foundation for the Advancement of Teaching*

A Guide for Leaders in Higher Education, Second Edition

Concepts, Competencies, and Tools

Brent D. Ruben, Richard De Lisi, and Ralph A. Gigliotti

Foreword by Jonathan Scott Holloway

"After an award-winning first edition, Brent Ruben, Richard De Lisi, and Ralph Gigliotti are back with a second edition of *A Guide for Leaders in Higher Education: Concepts, Competencies, and Tools*. This book could not come at a better time given the leadership challenges facing society like COVID-19 and issues of equity and social justice. The authors not only address higher education's role in meeting these challenges, but they expand their treatment of the book's core concepts and tools. As a result, they bridge theory and practice and underscore the communicative foundation of academic leadership in sophisticated fashion. The continuing importance of their work cannot be underestimated. It is a resource that all academic leaders need—and will thoroughly enjoy."
—*Gail T. Fairhusrt, Distinguished University Research Professor, University of Cincinnati*

"This book is unique in providing both frameworks and vital information needed for successful leadership in higher education. I recommend it to all of our department chairs and use it in our leadership development program. Coverage of essential topics such as the changing landscape of higher education, perspectives on leadership, and communication strategies for academic leaders makes this an essential resource for aspiring and current academic leaders."—*Eliza K. Pavalko, Vice Provost for Faculty and Academic Affairs; and Allen D. and Polly S. Grimshaw Professor of Sociology, University of Indiana, Bloomington*

Shared Leadership in Higher Education

A Framework and Models for Responding to a Changing World

Edited by Elizabeth M. Holcombe, Adrianna J. Kezar, Susan Elrod, and Judith A. Ramaley

Foreword by Nancy Cantor

Today's higher education challenges necessitate new forms of leadership. A volatile financial environment and the need for new business models and partnerships to address the impact of new technologies, changing demographics, and emerging societal needs, demand more effective and innovative forms of leadership. This book focusses on a leadership approach that has emerged as particularly effective for organizations facing complex challenges: shared leadership.

Rather than concentrating power and authority in an individual leader at the top of an organization, shared leadership involves multiple people influencing one another across varying levels and at different times. It is a flexible, collective, and nonhierarchical approach to leadership. Organizations that have implemented shared leadership have been better able to learn, innovate, perform, and adapt to the types of external challenges that campuses now face and that will continue to shape higher education in the future.

Intended as a resource for leaders at the highest levels such as presidents and provosts as well as midlevel leaders such as deans, directors, and department chairs, the book is also addressed to faculty and staff who are interested in collaborating with campus leaders on institutional decision-making or creating new change initiatives. It is intended to build capacity for shared leadership across institutions and for use in leadership courses and programs.

The Activist Academic

Engaged Scholarship for Resistance, Hope and Social Change

Colette Cann and
Eric DeMeulenaere

The Activist Academic serves as a guide for merging activism into academia. Following the journey of two academics, the book offers stories, frameworks, and methods for how scholars can marry their academic selves, involved in scholarship, teaching, and service, with their activist commitments to justice, while navigating the lived realities of raising families and navigating office politics. This volume invites academics across disciplines to enter into a dialogue about how to take knowledge to the streets.

"With all the humor, honesty, and humility that you'd expect in a conversation between friends, the dialogues between Cann and DeMeulenaere that span cover-to-cover animate the potential and the challenges of approaching research, teaching, and service as an activist academic. But be ready: Drawing deeply on theory and experience, this book will pull readers into the conversations, the inquiry, and the unavoidable demand as we dive into the unresolvable contradictions at the heart of being a professor committed to justice."—***Kevin Kumashiro***

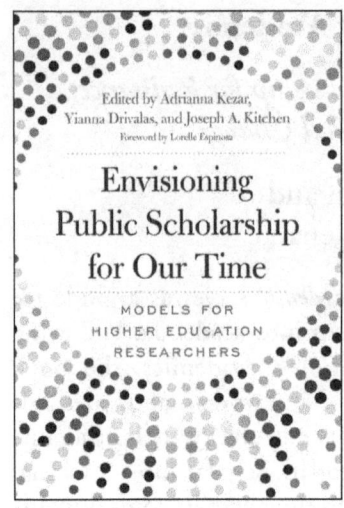

Envisioning Public Scholarship for Our Time

Models for Higher Education Researchers

Edited by Adrianna Kezar, Yianna Drivalas, and Joseph A. Kitchen

Foreword by Lorelle Espinosa

"As educators, this book reminds us of our shared responsibility to contribute, and more importantly, to be in service to the public good. Every emerging and current scholar should read this book with this question in mind: How will my work embody the definition of public scholarship as connected to a diverse democracy, equity, and an avenue for social justice? The answer has the potential to reshape how we conduct research and how we prepare future scholars."—***Tia McNair***, *Vice President for Diversity, Equity, and Student Success, Association of American Colleges and Universities*

"In *Envisioning Public Scholarship*, scholars offer accounts of why and how our social science research can be conducted for the democratic good. In the spirit of John Dewey's democratic ethics, scholars in these pages organize and operationalize democratic equity as both means and ends of research. 'Public scholarship' is not the old century's call for educating policymakers. Rather, this clarion's call is a millennial one—research as democratic activism, boldly presented and timely, indeed."—***Ana Martinez-Aleman***, *Professor, Lynch School of Education, Boston College*

Pitch Perfect

Communicating with Traditional and Social Media for Scholars, Researchers, and Academic Leaders

William Tyson

"In a time of growing scientific, technological, and ethical complexity in all aspects of our lives, the need for researchers and the academic community to reach out to the public has never been more important. *Pitch Perfect* provides easy access for the interested, but hesitant academic to get involved in some of the most important national and international conversations of our time. Bill Tyson has provided the tools for getting important information off the sidelines and into the national dialogue without compromising objectivity and scholarship."—***Thomas S. Litwin***, *Director, Clark Science Center, Smith College*

"In *Pitch Perfect*, author William Tyson provides a practical how-to guide for academics wanting to engage with traditional and new media. The book reads like a primer on the field of journalism with applications for academia. . . . The content of this book would be a good selection for academics who have valuable and newsworthy expertise or initiatives and are seeking to make these advances."—**NACADA Journal** *(National Academic Advising Association)*

Also available from Stylus

Publicly Engaged Scholars

Next Generation Engagement and the Future of Higher Education

Edited by Margaret A. Post, Elaine Ward, Nicholas V. Longo, and John Saltmarsh

Foreword by Timothy K. Eatman

Afterword by Peter Levine

"*Publicly Engaged Scholars* is an inspiring book that presents a renewed vision and pathway to advance participatory scholarship. The book provides a critical exploration of public engagement that is grounded in both a rich understanding of the historical context and future promise of socially responsible, impactful, and equitable scholarship. As a next generation scholar, the book challenges us to shape the future of engaged scholarship as active collaborators and change agents partnering with communities and within and across institutions of learning."—*Jomella Watson-Thompson*, 2014 Lynton Award Recipient; Assistant Professor; Associate Director of the Work Group for Community Health and Development, University of Kansas

"Grounded in the deep history of the engagement movement in higher education but spoken in the voices of its next generation, *Publicly Engaged Scholars* is both unflinching in its presentation of the challenges—personal, professional, political—facing those who seek to transform higher education for the greater good and hopeful in its demonstration of the persistence and adaptability of engaged scholarship. Anyone concerned about higher education's contribution to democracy should read it."—*Andrew J. Seligsohn*, President, Campus Compact

22883 Quicksilver Drive
Sterling, VA 20166-2102   Subscribe to our email alerts: www.Styluspub.com

PROCEEDINGS OF THE TWELFTH EUROPEAN CONFERENCE ON SOIL MECHANICS AND GEOTECHNICAL ENGINEERING/AMSTERDAM/NETHERLANDS/7-10 JUNE 1999

Geotechnical Engineering for Transportation Infrastructure

Theory and Practice, Planning and Design, Construction and Maintenance

Edited by
F.B.J. Barends, J. Lindenberg, H.J. Luger, L. de Quelerij & A. Verruijt

ADDITIONAL VOLUME

A.A. BALKEMA / ROTTERDAM / BROOKFIELD / 2000

Cover photo: Wim Ruigrok, Amsterdam.
A four-track railway tunnel construction in a strutted cofferdam trench under the town center of Best. From the year 2000 on trains will be passing this point. A new station will give entrance by elevators and stairs to the platforms. Building preparations were difficult. Seventy houses and barns had to be dismantled and a 600-thousand cubic meter of soil was removed. The 30 thousand inhabitants of Best used to quietness and tranquility, adjusted themselves to litter and noise. Early mornings, Best awakens when pile driving starts. Best is growing.

Authorization to photocopy items for internal or personal use, or the internal or personal use of specific clients, is granted by A.A.Balkema, Rotterdam, provided that the base fee of US$ 1.50 per copy, plus US$ 0.10 per page is paid directly to Copyright Clearance Center, 222 Rosewood Drive, Danvers, MA 01923, USA. For those organizations that have been granted a photocopy license by CCC, a separate system of payment has been arranged. The fee code for users of the Transactional Reporting Service is: 90 5809 134 1/00 US$ 1.50 + US$ 0.10.

Published by/Publié par
A.A.Balkema, P.O.Box 1675, 3000 BR Rotterdam, Netherlands
Fax: +31.10.413.5947; E-mail: balkema@balkema.nl; Internet site: www.balkema.nl
A.A.Balkema Publishers, Old Post Road, Brookfield, VT 05036-9704, USA
Fax: 802.276.3837; E-mail: info@ashgate.com

ISBN 90 5809 134 1

© 2000 A.A.Balkema, Rotterdam
Printed in the Netherlands/Imprimé aux Pays-Bas

Epilogue

Every time we use the telephone, turn on the light, pull the chain, open the tap, or surf on Internet, there is an underground motion, a subterrestrial impulse. Usually, we are unaware of the underground labyrinth that makes this possible, a true network with millions of kilometer of cables, wires and pipes. What we experience is the underground transport by car and train in tunnels and metro's, or parking and shopping under the surface. The underground is noticed particularly when things are wrong by fire, collapse or leakage. Furthermore, above-ground transport on tracks on which we can move by train, car, bicycle and by feet are supposedly well supported by stable earth bodies, and if not it is an unpleasant surprise. Mobile we want to be, more and more, but without disturbance.

The transportation infrastructure is at stake.

Therefore, the theme of the Amsterdam XIIth European Conference on Soil mechanics and Geotechnical Engineering 1999 has been dedicated to *geotechnical engineering for transportation infrastructure*, as mobility is now a common and appealing concern everywhere, with the focus on concrete targets, on what is thought valuable for politicians and policy makers, useful for the societies and future generations.

This volume – number four and last in row of the proceedings of the XIIth ECSMGE 1999 – contains three parts. In the first part *Reflection on the Conference* the Organizing Committee collected characteristic information, valuable for the record and to organizers of future events. The second part *Recapitulation of the conference* compiles summaries of sessions, workshops and specials that took place during the conference. And the third part deals with the results of the Workshop on *Information Technology in the Geotechnical Profession*. Finally, an ill-printed paper is reproduced and a *List of Registered Delegates to Amsterdam* is included.

As stated in the Preface of the proceedings and occasionally made clear during the conference, the real emphasis should be not on mobility but on interconnectivity in order to improve mobility in a wider sense. Interconnectivity becomes prominent in societies (globalization), in professional co-operation (multi-disciplinarily), in generic integration (the total approach) and in worldwide inspiration (Internet). In this regard the incorporation of the results of the Workshop on IT is justified, as it represents a modern way of transport, not physically but mentally, transport in a virtual world.

It is in this sense that the geotechnical profession could develop its creativity, significance and contribution, and prosper.

In the name of the editors,

Frans B.J. Barends
Chairman of the Scientific Committee

Contents

Reflection on the Conference   1
Consideration and Justification

Recapitulation of the Conference   21
Summaries of Sessions, Workshops and Specials

Information technology in the geotechnical profession
Full Papers of Workshop 4

Validation of computer programs in geotechnical design   57
A. Bond

The emergency of information technology – a state of practice report   65
K.R. Massarch

Information retrieval and communication   83
B. Rydell, A. Salomonson & J. Lindgren

Information technology applications in geotechnical education and vocational training   99
D.G. Toll

DESSYS, Geotechnical Design Decision System   113
M. van Veghel

Miscellaneous

Revised paper Volume 3: Aspects for dynamic compaction of saturated sand   129
P. Vuola & J. Hartikainen

List of registered delegates to the XIIth ECSMGE 1999 in Amsterdam   137

Reflection on the Conference
Consideration and Justification

The Organizing Committee

Louis de Quelerij   Jaap Lindenberg   Dirk Luger   Arnold Verruijt   Frans Barends

Freerk de Boer

Fred Jonker

John van der Kamp

Reflection on the Conference

LAST WORDS AT THE CLOSING CEREMONY OF THE XIIth ECSMGE 1999, AMSTERDAM

Louis de Quelerij, Chairman of the Organizing Committee

Dear ladies, friends and colleagues. The twelfth European Conference is coming to an end. During four days we have been participating in very interesting technical sessions and in a variety of social events. I hope you all enjoyed the happenings as much as I did.

Preparation of this conference took a long time. I will recall some of the key factors that have contributed to success. First, of all, we were very pleased that the Organizing Committee of the preceding (eleventh) conference in Copenhagen in 1995 offered us full co-operation to share their experiences and provide us with good advice.

One lesson from the Danish was that we should start at a very early stage in getting financial support. As good Dutch businessmen of course, we followed this advice and we were very pleased that we got financial support at an early stage from 5 main sponsors and 21 sponsors. By means of this power point projection, I like to thank all these sponsors, from which the main sponsors and some of the other sponsors contributed in addition significantly in many man-hours.

Apart from the financial side the contents of the technical sessions, in other words the quality of the scientific program formed a second key factor for success. As Organizing Committee we tried to focus the technical program on the application of new geotechnical developments in the different fields of infrastructure, rather than on the scientific tools themselves. We felt that interchange between academics and practical engineers (both consulting and contracting) has been very much encouraged in this way. The technical Program has been worked out into 5 main sessions, 10 discussion sessions and 8 workshops. In addition, we organized 5 keynote lectures, 12 special project lectures and 1 heritage lecture.

We appreciate the contribution of the Regional Conference Advisory Committee in 7 meetings for helping us selecting the discussions themes and more important selecting the right performers: the chairmen, discussion leaders, special lecturers and panelists. From the beginning William van Impe, Dick Parry, Michael Jamiolkowski and Jørgen Steenfelt helped us a lot. Later on, Heinz Brandl, François Schlosser and Michel Gambin supported us as well. Including helping us in French translations of the bulletins: merci beaucoup.

The third key factor was the quality of the performers, the people on stage. We think that the performers did an excellent job when I look back at the high quality of presentations and discussions. I would like to thank each of you for your high quality performance. Also the authors of in total 304 papers from which 26 outside of Europe (a record I think) are very much acknowledged. Almost 70 of them have presented their papers in very attractive poster sessions. I must say the preparation of the proceedings and in particular the uniform CD ROM quality was a very hard job for the Scientific Committee, with a major contribution from Frans Barends.

As we all know you as the delegates, the participants are the most important key factor for success of this conference. I can assure you that the gradual growth of the number of delegates kept the Organizing Committee in tension. Although we observed from previous conferences a correlation between the number of delegates and the number of papers and we are familiar with the backlash effect of an early bird period, it was not until the last moment that we felt happy with the total number of delegates. We are very pleased that in total 638 participants have been registered and about 100 accompanying persons. In total 58 countries worldwide were represented, of which 32 were European. In addition, 45 exhibitors from 14 countries were present at the Technical Exhibition.

The last key factor for conference success forms the quality of the Organizing Team. That is not only the core team as we called it but also the 35 members of the committees dealing with the scientific program, the technical secretaries of the sessions, the social events, the exhibition, the excursions and the accompanying persons program. I would like to thank all my colleagues for their support and of course for their performance in the pile-driving choir. Also the preparations from the Amsterdam Culture Factory for their creative input, the performance of Leoni Jansen and the services from the RAI Congress Center is appreciated. I would like to give my special thanks to John van der Kamp of the Conference Organization of the Dutch Royal Institute of Engineers. He did an extremely good job in this in his very controlled and often humorous way. In addition, also the help from the supporting group of students and assistants and secretaries at the desk and the two interpreters, are very much appreciated. Thank you for your support.

At the end of the technical conference program, we will enjoy the four Technical Excursions. Over 350 participants have registered for these excursions, visiting interesting and inspiring projects.

We also like to continue supporting further European co-operation in our profession. In this respect we are willing and able to exchange our experience and knowledge in the organization of conferences with the Czech Republic Geotechnical Society in helping them with the following thirteenth European Conference in Prague.

Finally, thank you for your attendance and active participation.

CLOSING SPEECH

Prof. Dr Heinz Brandl, ISSMGE Vice-President of Europe.

Ladies and Gentlemen,
The 12th European Conference on Soil Mechanics and Geotechnical Engineering approaches its end. This international event has been a full success in every respect:
– A high level of oral and written contributions and of the discussions. The proceedings will certainly find interested readers world-wide;
– Perfect organization;
– The excellent accompanying persons program;
Allow me some final remarks:

The move into the next century and the expected growth in transportation will require an efficient and high-performing infrastructure all over Europe. Safety and high-traffic flow is important, and any restrictions due to reconstruction and maintenance must be kept at a minimum.

It should be emphasized that building in unstable, heterogeneous, or soft soil and rock includes a significant higher calculation risk than is experienced by other branches of civil engineering. Consequently, proper design and construction require not only a firm theoretical knowledge but also comprehensive experience and engineering intuition. This involves in many cases design issues, which need to be ruled out during construction or even in the long-term according to the observational method.

It should be further emphasized that a so-called 100%-safety cannot be obtained in many cases of the geotechnical engineering (e.g. landslides). And this must be accepted by the public, the politicians and other decision-makers, and by lawyers. Action groups preventing the construction of buildings which are required for the infrastructure of modern society (e.g. highways, railways, power plants, waste disposal facilities) should be called to account for their egoistic or fundamental, or even anarchistic activities.

Regarding the international situation on geotechnical conferences and seminars, I am somewhat concerned that there are increasingly too many events taking place in the European countries, overlapping in time and topics. An extreme dynamic and speed has started to govern our lives. While weeks, days and hours still have the same length – at least objectively – more and more events are being packed into these time periods, leaving hardly enough possibility to go into real depth.

The alarm bells ought to ring in the 'event staccato': there is not enough time to carefully work on a project or prepare carefully contributions to conference, journals, etc. Consequently, professional conferences, seminars, etc. should be more concentrated and not 'diluted' regarding frequency and quality likewise, i.e. reduction of the number of events while increasing their quality and not repeating the same results again and again (very often just in a somewhat different packaging). Accordingly, our geotechnical community should focus on the International Conferences, the European Conferences, the Danube European Conferences, the Baltic See Conferences, and other regional events of the ISSMGE, but not support those numerous competitive events which spring up all over the place frequently with new names, e.g. Geotechnology, etc.

Now, coming to the end:

On behalf of ISSMGE, I would like to express our deepest thanks to the Netherlands Member Society, especially to Mr. Louis de Quelerij and his Organizing Committee, which did an excellent job. Many thanks also to Mr. Gerben Beetstra as Chairman of the Netherlands Member Society and special thanks to Prof. Frans Barends, Chairman of the Scientific Committee and always the overviewing personality. I would like to thank the exhibitors. Exhibitions always have been a source of interesting information. And finally many thanks to all of you for having contributed to make this conference such a success. I think that everybody will remember with pleasure these days in Amsterdam. I hope we shall meet again at the next European Conference of Soil Mechanics and Geotechnical Engineering in Prague, 2003.

Thank you very much and good-bye.

CONFERENCE STATISTICS AND FACTS TO REMEMBER

Frans Barends, Chairman Scientific Committee
John van der Kamp, KIVI Congress Office

Registration
The delegates to the XIIth European Conference on Soil Mechanics and Geotechnical Engineering 1999, Amsterdam, came from 32 European countries, and from 26 countries of other continents. The fact that the biannual Council Meeting of the ISSMGE was held in Amsterdam during the conference promoted the participation of delegates from outside Europe. The distribution of delegates' home countries is presented in the map, shown next.

The definite number of delegates is established at 638: in total 667 participants registered and 29 cancelled, 9 from Russia, 5 from Nigeria, 3 from Romania, 2 from Lithuania and 2 from Kazakhstan, and 8 form various other countries. A free registration was granted to 31 delegates, mainly from Eastern Europe, in accordance with a promise made during the 1995 contest for the conference hostage. 101 accompanying persons joined the social program. A strong reduced registration fee was granted to 19 students and 66 stand crewmembers, and a reduced fee to 27 exhibitors. 237 participants registered with the reduced early registration fee, and finally, 247 (about 37%) registered with the normal fee just a few weeks before the start of the conference. Since the

REGISTRATION PER COUNTRY

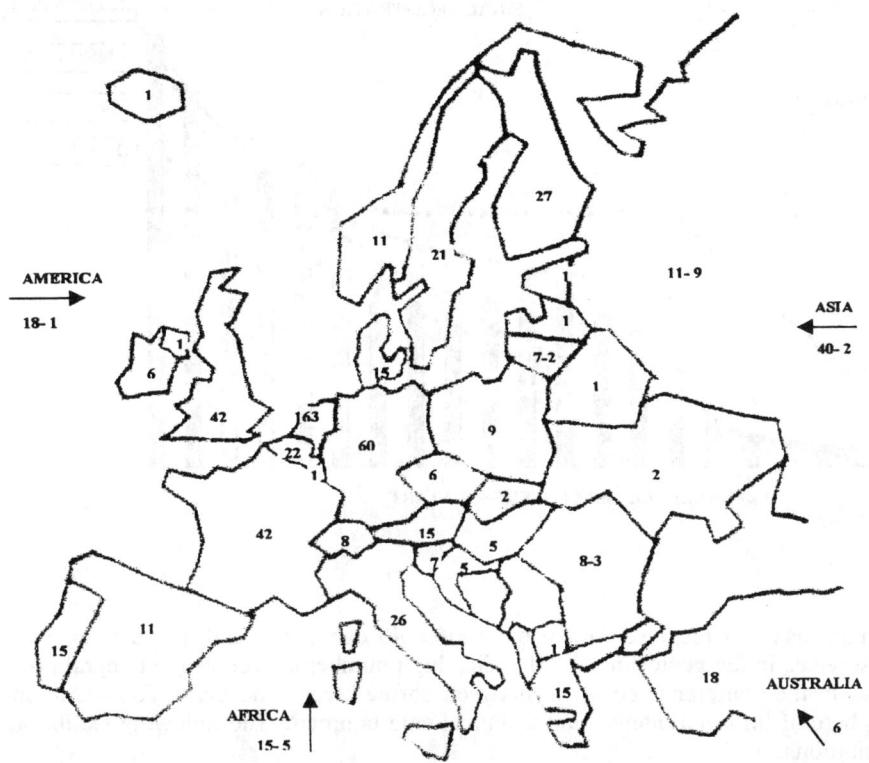

financial break even point was a number of 525 paying delegates, it tensed the nerves of the Organizing Committee until the last moment whether the conference could be a financial success. In comparison to previous European conferences (relevant data are shown in the graph by dotted lines). It does not seem important for the number of total delegates whether an early bird registration is chosen. The early bird registration has the convenience of providing liquidity in an early stage.

Excursions and social program
Four technical excursions planned directly after the closing ceremony allowed the delegates to say fare well to their new and old friends and colleagues in a relaxed and interesting ambience. Visits were organized and sponsored by several Dutch companies and Institutes to
– The Botlek railway tunnel at Rotterdam;
– IJburg land reclamation at Amsterdam;
– The Ketel-lake dredge disposal depot;
– Tram tunnel at The Hague.
Over 350 delegates took part in these events under fair Dutch weather conditions. The accompanying persons' program, the cultural conference banquet and/or the free public-transport week-ticket also served 101 accompanying persons. The Municipality of Amsterdam, the Ministry of Transport, Public Works and Water Management and GeoDelft have sponsored the informal get-together-party, the official welcome and cultural banquet. The Organizing Committee could therefore decide to include the remaining costs of all events and services into one single fee. This kept the delegates together and minimized social deviation.

6 Reflection on the Conference

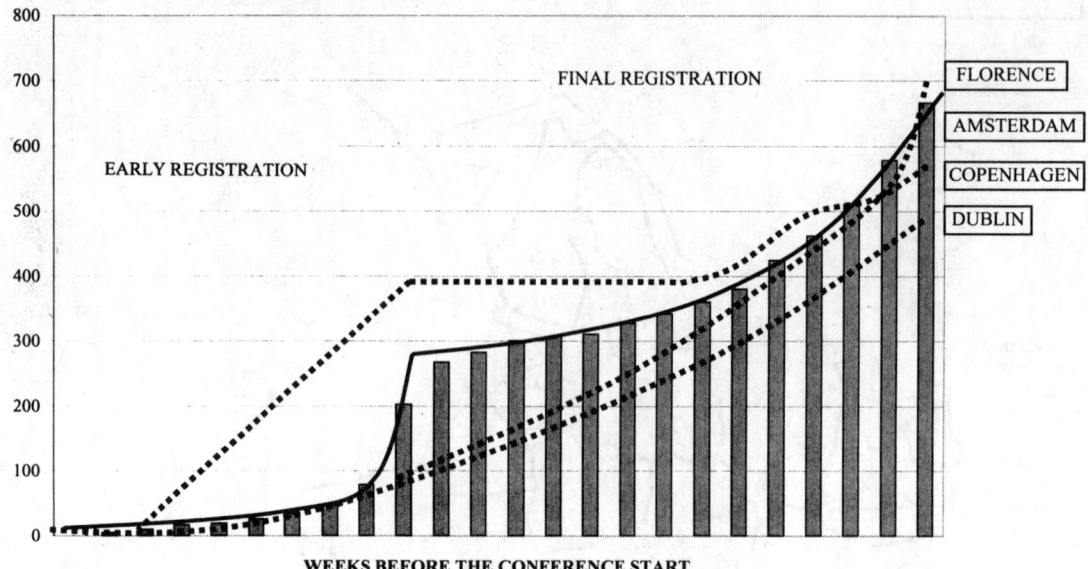

Exhibition
An overwhelming response was received for the exhibition. 45 companies and institutes exposed their facilities and services in the geotechnical field. This high number as well as the integral positioning of the stands in the conference center in between coffee breaks and discussion rooms indeed supported a sphere of informal interest and technical entertainment. The following exhibitors have organized exhibitions.
   Exhibitors at the XII European Conference of Soil Mechanics and Geotechnical Engineering at Amsterdam:
– Keller Grundbau, Offenbach, Germany
– PLAXIS, Rhoon, Netherlands
– TERRASOL, Montreuil, France
– Inpijn-Blokpoel, Son, Netherlands
– NAUE FASERTECHNIK, Leubbecke, Germany
– WILLE GEOTECHNIK, Goettingen, Germany
– Solexperts, Schwerzenbach, Switzerland
– Huesker Synthetics, Gescher, Germany
– Polyfelt, Linz, Autria
– Fugro, Leidschendam, Netherlands
– Fundex Companies, Oostburg, Netherlands
– STRUKTON, Maarsen, Netherlands
– GGU Gesellschaft für Grundbau und Umwelttechnik, Braunschweig, Germany
– Holland Railconsult, Utrecht, Netherlands
– DSI-DIWIDAG-Systems International, München, Germany
– GeoDelft, Delft, Netherlands
– Technosoft, Lochem, Netherlands
– GDS Instruments, Egham, United Kingdom
– Tensar International, Blackburn, Lancashire, United Kingdom
– A.P. van den Berg, Heereveen, Netherlands
– Boart Longyear, Etten-Leur, Netherlands

- Junttan Oy, Kuopio, Finland
- ID FOS RESEARCH EEIG, Geel, Belgium
- Rijkswaterstaat-DWW, Delft, Netherlands
- Loadtest, Gainesville, Fl, United States of America
- EMAP Construct, London, United Kingdom
- ProfilARBED-ISPC, Esch/Alzette, Luxembourg
- Geonor, Oslo, Norway
- 'NEDEXIMPO'/HSP Hoesch Spundwand und Profil, Amsterdam, Netherlands
- SISGEO, Segrate (MI), Italy
- Pagani Geotechnical Equipment, Calendasco, Italy
- GEOTECH, Askim (Gothenborg), Sweden
- Netherlands Institue of Applied GeoScience TNO-NGC, Delft, Netherlands
- Terre Armee – Reinforced Earth, Breda, Netherlands
- Geotechnics Holland, Amsterdam, Netherlands
- TNO-Bouw/DIANA Analysis, Delft, Netherlands
- IHC Hydrohammer, Kinderdijk, Netherlands
- Geomil Equipment, Alphen a/d Rijn, Netherlands
- MICRONIC, Berlin, Germany
- BAUER Spezialtiefbau, Schrobenhausen, Germany
- Delft University of Technology, Delft, Netherlands
- APAGEO SEGELM, Magny les Hameaux, France
- John Wiley & Sons, Chichester, Sussex, United Kingdom
- Koenders Instruments, Almere, Netherlands
- A.A. Balkema, Rotterdam, Netherlands

Sponsors
A conference of the shape of the XIIth ECSMGE cannot be realized without the invaluable support by sponsors, who contributed in terms of finance and non the less in an impressive number of free hours by the organizing crew, i.e. their companies allowed them to spend working time. The Organizing Committee decided in early stage to invite a limited number of main sponsors who were asked to cover the missing estimated budget – about 15% of the total estimate – and thereafter to give opportunity to other sponsors allowing them some privileges at the conference. This approach was successful. The five main sponsors reacted almost instantly positively, and the list of other sponsors was growing steadily.
Main sponsors:
- *A.P.v.d. Berg*
- *GeoDelft*
- *Fugro*
- *Holland Railconsult*
- *Ministry of Transport, Public Works and Water Management*

and other sponsors:
- *ABT Adviesbureau voor Bouwtechniek bv*
- *ARCADIS Bouw/infra B.V.*
- *Ballast Nedam N.V.*
- *B.V. AdviesbureauTjaden voor Technisch Bodemonderzoek*
- *CUR/COB*

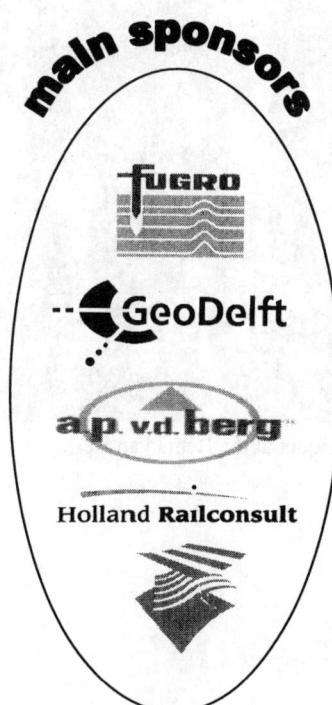

8 Reflection on the Conference

- De Ruiter Boringen en Bemalingen B.V.
- DHV Milieu en Infrastructuur BV
- Haskoning Ingenieurs- en Architectenbureau
- HBW, HWZ, HAM-VOW bv
- infraneth
- Ingenieursbureau van Gemeentewerken Rotterdam
- MOS GRONDMECHANICA B.V.
- Multiconsult BV
- Nederlands Instituut voor Toegepaste Geowetenschappen TNO
- Omegam
- Oranjewoud infragroep
- Raadgevend Ingenieursbureau Inpijn-Blokpoel
- Royal Boskalis Westminster nv
- Strukton
- TNO-Bouw
- VAGWW/NVAF

And all others, who contributed, we all owe them a cordial 'thank you'.

Banquet at the Beurs van Berlage

Reflection on the Conference

XIIth European Conference on Soil Mechanics and Geotechnical Engineering

Amsterdam, The Netherlands
7-10 June 1999

WHO IS WHO?*

A. Regional Conference Advisory Committee
B. Organising Committee
C. International Scientific Committee
D. Exhibition Committee
E. Social Committee
F. Accompanying Person's Committee
G. PR Committee
H. Technical Excursions Committee

I. Scientific Committee
J. Chairman
K. Co-chairman
L. Keynote Lecturer
M. Special Project Lecturer
N. Discussion Leader
O. Panelist
P. Secretary

* If photo available

D. Adam
(Austria)
M

C.G. Alén
(Sweden)
O

E.E. Alonso
(Spain)
C, J, M

A.G. Anagnostopoulos
(Greece)
C

K.H. Andersen
(Norway)
C, O

O. Arkima
(Finland)
M

K.J. Bakker
(The Netherlands)
P

F.B.J. Barends
(The Netherlands)
A, B, I, O, P

E. Barends-Zoabi
(The Netherlands)
F

G. Barla
(Italy)
J

E.N. Bellendir
(Russia)
O

P.E. Bengtsson
(Sweden)
O

P. van den Berg
(The Netherlands)
P

B. Berggren
(Sweden)
N

F. de Boer
(The Netherlands)
A, B, H

M. Boulon
(France)
O

H. Brandl
(Austria)
A

M. Van den Broeck
(Belgium)
N

10 Reflection on the Conference

G. Calabresi
(Italy)
C, N

S. Cavounidis
(Greece)
J

S.F. Çinioğlu
(Turkey)
J

J.B. Clausen
(Denmark)
O

F. de Cock
(Belgium)
J

A.G. Correia
(Portugal)
J

R.M. Correia
(Portugal)
M

V. Cuellar
(Spain)
O

H. Denver
(Denmark)
N

N. Derksema
(The Netherlands)
B

J.D. van Duijvenbode
(The Netherlands)
E, P

M. Dysli
(Switzerland)
O

P. Egger
(Switzerland)
O

C. Erichsen
(Germany)
O

E.R. Farrell
(Ireland)
O

J. Feda
(Czech Republic)
C

A.M. Gaberc
(Slovenia)
O

M. Gambin
(France)
B

H.C. van de Graaf
(The Netherlands)
E

J.H. Gravgaard
(Denmark)
M

G. Greschik
(Hungary)
J

G. Hannink
(The Netherlands)
H, P

J. Hartlén
(Sweden)
C, J

G.T. Houlsby
(United Kingdom)
L

L. Huizer
(The Netherlands)
M

V.A. Ilyichev
(Russia)
C, K

W.F. van Impe
(Belgium)
A

E. Imre
(Hungary)
O

M. Jamiolkowski
(Italy)
J

R. Jardine
(United Kingdom)
N

F. Jonker
(The Netherlands)
B

J.M. van der Kamp
(The Netherlands)
B, D, E, G

K. Karlsrud
(Norway)
L

M. Kavvadas
(Greece)
M

H.J. Kolk
(The Netherlands)
P

I.L. Kolk-Eileraas
(The Netherlands)
F

Reflection on the Conference

N. Krebs Ovesen
(Denmark)
J

T.J. Kvalstad
(Norway)
O

E. Kwast
(The Netherlands)
D

S. Lacasse
(Norway)
J

E. Leca
(France)
L

J. Lindenberg
(The Netherlands)
B, I

L.F. Linney
(United Kingdom)
M

A. Loizos
(Greece)
N

J.A. Lord
(United Kingdom)
N

P. Lubking
(The Netherlands)
E

E.Th. Luger-van Es
(The Netherlands)
F

H.J. Luger
(The Netherlands)
B, I, O

J. Maertens
(Belgium)
M

L. Maertens
(Belgium)
C, K

J.-P. Magnan
(France)
C, L

R.J. Mair
(United Kingdom)
N

E. Maranha das Neves
(Portugal)
C

A. Marcu
(Romania)
C, O

B. Marić
(Croatia)
C

L. Martak
(Austria)
O

T.W. McNeilan
(USA)
M

M.T. van der Meer
(The Netherlands)
P

I.K. Mihalis
(Greece)
O

J.D. Nieuwenhuis
(The Netherlands)
C

M. Nussbaumer
(Germany)
C, N

A.-L. Öberg-Högstra
(Sweden)
O

T. Orr
(Ireland)
C, J

H.M.A. Pachen
(The Netherlands)
P

D. Parry
(United Kingdom)
A

G.W. Plant
(Hong Kong)
M

L. de Quelerij
(The Netherlands)
A, B, E, I

G.H. van Raalte
(The Netherlands)
O

H. Rathmayer
(Finland)
C, O

G. Roques
(France)
O

B. Rydell
(Sweden)
J

B. Rymsza
(Poland)
O

12 Reflection on the Conference

L.E.B. Saathof	C. Sagaseta	G. Scarpelli	F. Schlosser	H.R. Schneider	P. Sêco e Pinto
(The Netherlands)	(Spain)	(Italy)	(France)	(Switzerland)	(Portugal)
P	O	O	A, J	C, O	K

E.P.T. Smits	U. Smoltczyk	J.S. Steenfelt	M. Stocker	R. Szepesházi	H. Thurner
(The Netherlands)	(Germany)	(Denmark)	(Germany)	(Hungary)	(Sweden)
H	J	B, N	J	O	O

E. Togrol	M. Topolnicki	P. Turček	J.A. van Twillert	I. Vähäaho	A. Valkeisenmäki
(Turkey)	(Poland)	(Slovakia)	(The Netherlands)	(Finland)	(Finland)
C, J	O	J	P	O	O

A.A.M. Venmans	P.A. Vermeer	A. Verruijt	W.S.A. Verruijt-Hoogewerf	G.T. Visser	E. Welling
(The Netherlands)	(Germany)	(The Netherlands)	(The Netherlands)	(The Netherlands)	(The Netherlands)
P	K	A, B, C, I, L	F	H	D

W. Wittke	R.F. Woldringh	W. Wolski	D. Znidarčić
(Germany)	(The Netherlands)	(Poland)	(Croatia)
M	O	C, K	O

Special facilities
As a follow up of the facility so successfully offered at the Copenhagen Conference, a computer support at the information desk for text processing for transparency and beamer presentation, and for flyers and ad hoc meetings proved to be very helpful and efficient. The full time student support of the *Praktische Studie,* Faculty of Civil Engineering, Technical University of Delft, in the audience rooms for a smooth performance of the presentations was invaluable.

Posters
In order to stimulate all scientists en engineers to participate, learn and exchange knowledge and experience it was decided to promote the poster sessions by asking every author personally to also prepare a poster and win the poster award of 1000 Euro. This idea has proven to be successful, since almost 100 posters, being 35% of all accepted papers, have been presented during a very well visited session. Yet, the question remains whether such a conference should be organized with the emphasis on a minimum of parallel sessions or a maximum, and so on an integral exchange or on a specific exchange. For the Amsterdam conference the Organizing Committee judged an integral exchange to be the better within the present day vision of the engineering profession.

Proceedings
The Scientific Committee, being aware that all papers (should) have been evaluated by the national geotechnical societies, did not accept 2 papers, one paper has been completely 'translated' by the editors, 51 French titles and abstracts have been formulated and added by the editors, and one English title and abstract. For the French text in the bulletins, program and parts of the proceedings much help and assistance was obtained from Henk van der Graaf, Michel Gambin and William van Impe, and for the English text from Dick Parry.

The XIIth ECSMGE proceedings is available in three conventional volumes of 2248 pages with – for the first time – all papers also on CD-ROM, with all its advantages. The combination of a book and a CD required earlier deadlines for authors and a special effort for the editors. The Scientific Committee received several computer viruses and all possible formats of disks, documents and pictures. Particularly the electronic mailing of text was mostly perfect, the electronic mailing of pictures, included in the written contributions or not, was sometimes a disaster.

The specifications for sending pictures should be defined more strictly. It is advised to e-mail pictures only in B&W and by JPG or (zipped) EPS format, maps in GIF format, and tables and graphs only when being produced with the textprocessor or being composed as vectorplot. The annotation in pictures and graphs should be superimposed by the textprocessor in a two point smaller font than the normal text. For authors not able to create the paper in the prescribed manner, detailed instructions should be made available.

The distribution of papers expected and received from the various countries is compiled in the table on page 14 and 15, together with the number of registered delegates.

What can be observed, is a remarkable increase of accepted conference papers. A reason for this achievement is the fact that in the conference special attention has been given to important large European projects, presented in the main sessions. The purpose was to emphasize more the engineering and construction aspects of the conference theme. It was a successful action.

The high number of registration waivers (31) and reductions (19) mainly for eastern European delegates and students was, after all, a wise action. Was it not, one should observe a tendency of decreasing European delegates, which is only partly due to economic differences of western and eastern European countries.

Lessons to be learned
– The early-stage experience transfer from previous conference organizations is very valuable.
– The Regional Conference Advisory Committee proved very helpful.
– The good team spirit in the Organizing and Scientific Committee proved to be invaluable. The time for the organization – about four years – is indeed required.

14 Reflection on the Conference

Country	Local country name	Registered delegates	Papers estimated originally	Received Papers abstracts	Papers complete
Austria	Österreich	15	6	5	5
Belarus	Belarus	1			
Belgium	België/Belgique	22		11	10
Bulgaria	Bulgaria		2	3	1
Croatia	Hrvatska	5	3	2	2
Czech & Slovakia Republics	Ceská & Slovenska Republici	8	3	5	5
Denmark	Danmark	15	10	19	13
Estonia	Eesti	1	2	2	1
Finland	Suomi	27	8	16	15
France	France	42	19	32	27
Germany	Deutchland	60	23	40	29
Greece	Ellada	15	7	11	10
Hungary	Magyarorszag	5	5	6	1
Iceland	Ísland	1	2		
Ireland	Eire	6	4	6	5
Italy	Italia	26	11	13	10
Latvia	Latvija	1	2		
Lithuania	Lietuva	7	2	7	4
Luxembourg	Luxembourg	3	9		
Macedonia	Makedonija	1			
Netherlands	Nederland (host)	163	20	38	30
Northern Ireland	Ulster	1			
Norway	Norge	11	8	12	11
Poland	Polska	9	7	9	7
Portugal	Portugal	15	4	6	6
Romania	România	8	4	9	9
Russia	Rossijskaja Feduracija	11	9	15	13
Slovenia	Slovenija	7	2	2	1
Spain	España	11	9	16	15
Sweden	Sverige	21	14	19	18
Switzerland	Schweiz/Suisse	8	10	9	6
Turkey	Türkiye	18	6	4	4
Ukraine	Ukraïna	2		2	2
United Kingdom	United Kingdom	42	21	29	23
Yugoslavia	Jugoslavija		2	2	2
Reserve			3		
special papers			18	Included per country	
SUBTOTAL EUROPE			256	350	285
Algeria		1		2	1
Australia		6		3	1
Bolivia		1			
Brazil		2		3	
Cameroon		1			
Canada		3			
China		4		1	1
Colombia		1			
Egypt		1		2	2
Hong Kong		1		3	3
India		6		6	3
Iran				1	1

Country	Registered delegates	Papers estimated originally	Received	
			Papers abstracts	Papers complete
Israel	1			
Japan	18		7	5
Kazakhstan	4		4	2
Korea, Republic	2			
Laos	1			
Lebanon	1			
Malaysia	1			
Mexico	1			
Morocco	2			
Nigeria	8			
Saudi Arabia	1			
Singapore	2		2	2
South Africa	1			
Thailand	2			
Tunisia	1			
United Arab Emirates	1		1	1
USA	10		4	4
SUBTOTAL NON-EUROPE		25	39	26
TOTAL	667	281	389	311

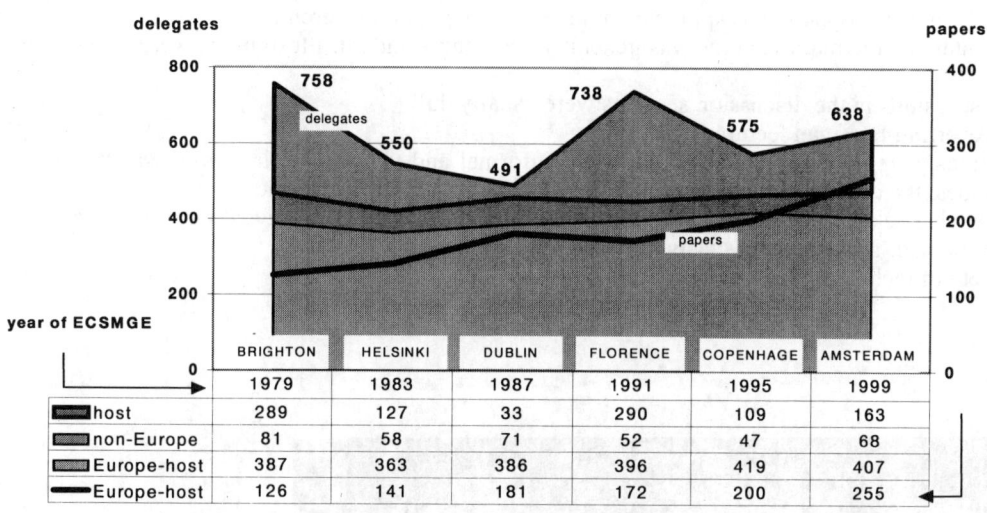

PARTICIPATION TO EUROPEAN CONFERENCES

year of ECSMGE	BRIGHTON 1979	HELSINKI 1983	DUBLIN 1987	FLORENCE 1991	COPENHAGE 1995	AMSTERDAM 1999
host	289	127	33	290	109	163
non-Europe	81	58	71	52	47	68
Europe-host	387	363	386	396	419	407
Europe-host	126	141	181	172	200	255

Lesson to be learned (continued):
- Quite some registered eastern European delegates didn't show up, without any message. This loaded the variable costs unnecessarily.
- Only one invited key person could not come.
- Students were enthusiastic about the reduced fee.
- An acceptable fee reduction for the exhibition crews was not foreseen.
- Some hotels were too far and public transport took then too much time.

Reflection on the Conference

- The organization was strict and at the same time flexible, allowing changes and inclusions in the program when necessary.
- The choice of a non-commercial congress office was perfect.
- The production of the proceedings was cumbersome and took much energy and time, mainly due to the fact that the XIIth ECSMGE 1999 was the Guinea pig for a combined production of proceedings and CD-ROM.
- A noticeable number of authors didn't produce the paper according to the abstract.
- Deadlines are always surpassed; the winner (with an important paper) did by 2.5 month!
- Author's instructions are not carefully read.
- Sending electronic pictures (photographs, tables, and maps) was a great problem, since there is not yet a generally known best way. We received several electronic viruses!
- 51 (English writing) authors didn't produce the French abstract and title.
- 1 (French writing) author didn't produce the English abstract and title.
- A short standard keyword list for paper characterization was not available.
- The planning of the production of the bulletins was underestimated.
- The exhibition divided over two floors gave some discrimination.
- The poster session was too condensed both in place and time. Authors would wish more opportunity to present their work.
- Since only one poster session was planned, poster presenters, being posted at their own, could not join the vivid discussions at some other poster presentations during the session, unfortunately. This was not foreseen.
- The best-poster award worked well.
- The all-in price was well received.
- The student help-desk for using the projection instruments prevented hampering presentations.
- The beamers for computer presentations were well used (4 beamers in use).
- The informal atmosphere was appreciated; the new-style opening ceremony set the trend.
- The content of the main lectures was generally speaking standard; illustrations were often very good.
- The discussions in the discussion sessions were usually dull.
- The Keverling Buisman lecture was eminent.
- The atmosphere at the congress banquet was informal and very nice. The music was thrilling, sometimes, the quality of the food was somewhat disappointing, however.
- The Dutch Pocket Cheese Compressibility Test (DPCCT) was a nice event and with it a surprizing utensil; the cheese was tasty, not so durable.
- A variety of technical tours in a last Thursday afternoon is the best.

A Dutch Welcome

The Netherlands is known by soft soil, wooden shoes, tulips, pile driving and cheese. During the contest of the hosting country for the XIIth European Conference on Soil Mechanics and Geotechnical Engineering the voting members of the European Council were attracted by the Dutch dele-

Heilied

Hoog op mannen die nog heien kannen
Hoog op, slaat hem op zijn kop
Laag neer,
Trekt met het touwtje
We heien alweer

Al die slagen, kan hij verdragen
Hoog op, de paal op zijn kop (2x)

Dat is een, dat is twee
We heien al mee
Dat is drie, dat is vier
Meteen komt de baas met een glaasje bier
Dat is vijf, dat is zes
Meteen komt hij met een volle fles

Al die slagen, kan hij verdragen
Hoog op, de paal op zijn kop (2x)
Dat is zeven, dat is acht
Mannen wacht
Dat is negen, dat is tien
Nu zullen we nog wel eens zien
Dat is elf, dat is twelf
Ziet hij staat vanzelf

Al die slagen, kan hij verdragen
Hoog op, de paal op zijn kop (2x)

Drive song

High up men, if drive you can
High up, beat its head
Low then
Pull the rope
We're driving again

All those blows, and down it goes
High up, the pile on its butt (2x)

That makes one, that makes two
We join and we do
That makes three, that makes four
The boss will come, with ale in store
That makes five, that makes six
Soon he will come with a good old mix

All those blows, and down it goes
High up, the pile on its butt (2x)
That makes seven, that makes eight
Men wait, men wait
That makes nine, that makes ten
Now we shall see again
That makes eleven, that makes twelve
Look it stands by itself

All those blows, and down it goes
High up, the pile on its butt (2x)

gation by a souvenir broach of wooden shoes and a tag of tulips. At the conference opening ceremony a famous old pile driving song directed by Leoni Jansen and executed by the Organizing and Scientific Committee emphasized the way Holland was founded in the last centuries. Soft soil and cheese were integrated in the form of a new pocket compressibility meter, which the delegates could try out with Dutch fresh cheese. The poster session was stimulated with a poster award of 1000 Euro and the Honorary Senior team: Prof. Bram van Weele, Bert de Leeuw, Kees Joustra and Wim Heijnen selected the winner.

It was the purpose of the Organizing Committee to introduce some culture, self-doing and light competition and by doing so enhancing the informal aura of the event.

The Rat Parable
The question by Leoni Jansen, who led the cultural opening of the conference, to Prof. Dr H. Brandl: *Are you satisfied with the European Technical Committees?* was answered by Prof. Brandl as follows.

This is a rather delicate question. In analogy to global and partial safety factors in soil mechanics, I would say that, over all, I am satisfied. But in detail, the partial working factor is sometimes below $F = 1$. This could perhaps be compared to some recent results in animal behaviour research.

Whenever a group of three laboratory rats, which are very intelligent animals, were put together in a situation where they had to work in order to get their food, only one of them did the work while the others simply waited without doing anything, but then they shared the consumption of the meal. When, subsequently, only working rats were put together, the pattern repeated itself; again only one worked and two stand-by rats were waiting for the results.

This is obviously a natural law that also applies to human beings. It seems to me that we have a similar situation in the European Technical Committees. There certainly are members who put in an extraordinary high amount of work, while others contribute only in a minimal way, and sometimes not at all.

The Dutch Pocket Cheese Compressibility Test
(© Ruud Termaat, Jan-Dirk van Duijvenbode en Piet van der Velde)
The Ministry of Public Works and Water Management (RWS-DWW) and Fugro developed a Dutch Pocket Cheese Compressibility Test for the XIIth European Conference on Soil Mechanics

and Geotechnical Engineering in Amsterdam. It was presented during the opening session and every participant got a special cheese test kit with the Dutch Pocket Cheese Compressibility Test, three different types of cheese and a contest manual. They were asked to measure the E-modulus (Young's modulus) of Dutch fresh cheese at their hotel room.

The side and cross view of the apparatus are shown in the next figure. The plate and spring are connected with the top bucket. The force and displacement parameters D and H are registered at the outside scale on the bottom bucket. In this example D = 35mm and H = 18mm.

The two steps: calibration of the DPCCT and the testing of the cheese are explained next. These two steps can also been seen at the photo.

Step 1 Measure the spring constant of your DPCCT. To calibrate the Dutch Pocket Cheese Compressibility Test one measures the distance [mm] with a known weight [kg] at the top. Calculate the DPCCT Constant in mm/kPa with: C_{DPCCT} = Factor.$(D_0 - D_1)/(W_1 - W_0)$ [mm/kPa] (In this case the value of the Factor is 0,04).

Step 2 Measure the E-modulus of the fresh cheese sample. Install a 2x2x2 cm^3 block of fresh cheese in the Dutch Pocket Cheese Compressibility Test and measure the (D)istance and (H)eigth of the cheese sample in mm at different load steps (by hand pressure). Calculate the E-modulus

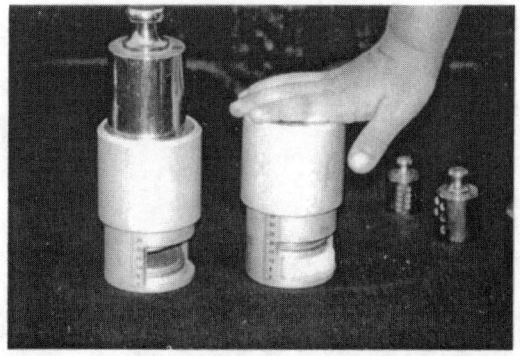
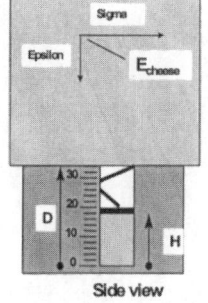
Side view
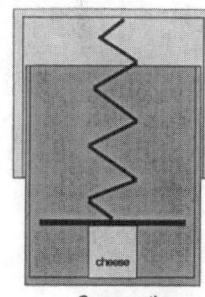
Cross section

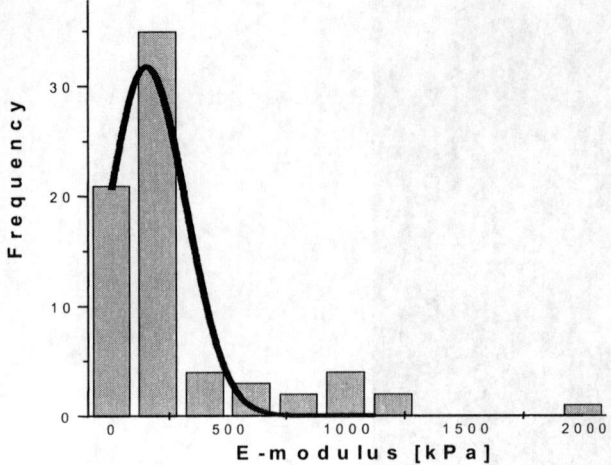

with: Epsilon = $(H_0 - H_1) / H_0$ [-]; Sigma = $((D_0 - D_1) - (H_0 - H_1))/C_{DPCCT}$ [kPa]; E-modulus = Sigma/Epsilon [kPa] = . . . , . . kPa.

The results are shown in the histogram. During the cultural banquet a life plate bearing test with 3 big pieces of cheese and the weight of Heinz Brandl was performed. An E-modulus of 107 kPa was measured.

Recapitulation of the Conference

Summaries of Sessions, Workshops and Specials
Session Secretaries (Netherlands)
Arjen Venmans
Benno Koehorst
Martin van der Meer
Sander Eijgenraam
Geerhard Hannink
Peter van den Berg
Harry Kolk
Maarten Smits
Harry Pachen
Jan-Aart van Twillet
Jaap Deutekom
Klaas-Jan Bakker
Herke Stuit
Koos Saathof
Bram van Weele
Frans Barends
Fred Jonker
Jaap Lindenberg
Wim Heijnen
Louis de Quelerij
Maurice van Veghel
Arnold Verruijt
Dirk Luger
Freerk de Boer
Joek Puechen
Fred Jonker

Main session 1 – General aspects of transportation infrastructure

Chairman: Prof. H. Brandl (Austria)
Co-chairman: Prof. P. Sêco e Pinto (Portugal)
Secretary: A.A.M. Venmans (Netherlands)

After the enthusiastic community singing by the audience in the official opening of the conference, Prof. H. Brandl opened as chairman the first main session. He pointed to the increased mobility between cities in Europe, calling for the development of a highly efficient traffic network. In the design and execution of these infrastructure works, safety for workers and environment will become more important than before. Also, ever more stringent demands are being set for minimization of maintenance. Finally, tomorrow's increasingly structurally sensitive structures will need to be built in areas considered marginal or unsuitable only yesterday.

Dr. K. Karlsrud (Norway) addressed these very topics in his keynote lecture. Based on his experience at the Norwegian Geotechnical Institute, he could identify the following challenges for the geotechnical profession:
– A need for the development of new and innovative solutions in order to reduce costs and achieve safety and quality of the works
– A need to the further development of construction methods that could limit displacements and associated malfunctioning and damage to neighboring structures and utilities;
– The design of infrastructure on very soft ground, aiming at further development of ground improvement techniques;
– The full integration of environmental aspects into all infrastructure developments.

Dr. K. Karlsrud clearly illustrated these challenges by a number of recent case histories from railway and road projects in Norway, details of which are given in the first volume of the proceedings. The main concern for the future is the development of an open dialogue with clients, using professional societies as forums for debate and lobbying.

In the closure of the session, Prof. P. Sêco e Pinto expressed his agreement with the statements made by Dr. K. Karlsrud, and awarded his keynote with four E's: Extravagant, Excellent, Exciting and Elegant.

Discussion session 1.1 – Geotechnical impact of transportation infrastructure

Chairman: Dr. S. Cavounidis (Greece)
Discussion leader: Prof. G. Calabresi (Italy)
Secretary: B.A.N. Koehorst (Netherlands)

In his introduction, chairman Prof. G. Calabresi showed that the title of this session can be understood in three different ways:
– The impact of geotechnics on the design,
– The impact of the design on geotechnics,
– The geotechnical impact of transportation on infrastructure.

In the presentations of the panelists the different points of view on the theme have been shown.

The first presentation was held by I. Vahaaho (Finland) on possibilities of tunneling engineering. He showed the different effects of tunneling in comparison to on-ground structures. The effects on costs, functions and social aspects are evident. In the design phase of a project a complete list of functions, the structure should perform, should be made.

Dr. A.-L. Öberg-Högsta (Sweden) emphasized in her presentation that geotechnics should play an important role in planning and design. In the near future this will be more important because controlling time-schedules and finances become more and more important for project managers. Therefore, the geotechnical discipline should be present in the very beginning of the project i.e. in the planning phase and in the design phase in order to prevent major disasters in exceeding the limits in time and finances due to a lack of attention to the geotechnical circumstances. Risk analysis is an important tool to determine and to deal with the uncertainties. Finally, Dr. Öberg-Högsta put the attention on the new guideline *Investigation of the stability of a slope*.

I. Mihalis (Greece) showed in his presentation that it is possible to stabilize the highly weathered rock in Athens in the metro-project. A jet-grouting method was used to minimizing the settlements and failures, successfully. This presentation showed the importance of a good technical design on controlling the costs and time.

Prof. A. Marcu (Romania) showed that large settlements and loss of stability of slopes could damage the road and railway embankments build on collapsible subsoil in Romania. Most of the time the deformations occur suddenly. The main course is a decrease of compressibility and strength due to a rise of the water content in such layers.

Prof. G. Calabresi (Italy) showed an example of a delicate tunneling project near the foundation of Castel San Angelo, a very important monument in Rome. To predict soil displacements both empirical and finite element analyses have been used. The risk of damage to the castle is determined using a method presented by Prof. J. Burland in 1995 in the first Conference on earthquake geotechnical engineering in Tokyo.

In the discussion with the audience the following remarks were made:
- Correct predictions of soil movement are very important;
- For this prediction non-linear behavior should be considered carefully;
- New techniques will increase the quality of the prediction, but one must not rely on new techniques only; the experience gained in the past should not be forgotten.

Discussion Session 1.2 – Dealing with uncertainties

Chairman: Prof. P. Turcek (Slovakia)
Discussion Leader: Dr. H. Denver (Denmark)
Secretary: M.T. van der Meer (Netherlands)

Prof. H. Denver noted that 28 papers have been submitted, from which only 50% involved statistics. In his presentation, he emphasized the importance of dealing with censored data. Dr. C. Alén (Sweden) showed a provocative example, to show that the selection of basic combinations is essential, in order to avoid a check of all possible combinations. H.J. Bakker (Netherlands) presented Prof. H. Vrijling papers, and stated that there are many different kinds of uncertainties, which should be dealt with in different ways. Dr. C. Sagesata (Spain) showed that in order to avoid overconservative, impossible constraints for sensitive structures, the observational method could be of very good use. Dr. H. Schneider (Switzerland) pointed out that the main uncertainties in a design problem are related to the soil itself, and presented some practical methods to obtain conservative selected mean values. The discussion was modest. Prof. P. Turcek (Slovakia) closed the session, and concluded that dealing with uncertainties is an important challenge in geotechnical engineering.

Main Session 2 – Harbours and Waterways

Chairman: Prof. W.F. van Impe (Belgium)
Co-chairman: Prof. W. Wolski (Poland)
Secretary: A.A. Eigenraam (Netherlands)

Session 2 of the XIIth European Conference on Soil Mechanics and Geotechnical Engineering consisted of 3 three presentations.

The first presentation by R.M. Correia (Portugal) of the National Laboratory of Civil Engineering in Portugal, was entitled 'The Vasco de Gama Bridge over the river Tagus in Lisbon – Main Geotechnical aspects'. After having mentioned the main characteristics of the bridge itself the geological and geotechnical aspects were pointed out. The soil profile at the crossing site was found to be composed of alluvial deposits and Plio-Pleistocene materials. In order to obtain the characteristics of the soil layers many boreholes, standard penetration tests and piezocone penetration tests were carried out. In addition some other field tests, such as vane tests, crosshole tests and seismic cone penetration tests were performed. Besides the field investigation many laboratory tests were carried out, mostly identification tests, oedometer tests and triaxial tests.

Since the area was considered to be susceptible to liquefaction the SPT, CPT and seismic tests were used to locate the zones that are vulnerable to liquefaction. In reference to the design of the pile foundations these zones were considered to be unsuitable for mobilizing any lateral resistance.

The bridge was founded on large diameter bored piles (1.8 m up to 2.2 m). In order to be able to calibrate the design parameters for a pile foundation and to optimize the pile length static (horizontal and vertical) and dynamic load tests were carried out. The static load tests showed that, in general, the predicted vertical failure loads were somewhat higher than the measured failure loads. This was caused by the overestimated shaft friction values.

The second presentation was on the subject of quay walls of the Rotterdam Harbour and was highlighted by Prof. A.F. van Tol (Netherlands) of Delft University of Technology/Rotterdam Public Works.

Due to the increasing dimensions of the ships the design and construction of the quay walls in the harbour of Rotterdam have been adapted over the last three decades and will develop furthermore in the future. The current status of design is a result of a comparative study that was carried out after having observed substantial displacements of the first 'EKOM' quay wall. The design concept of this 'EKOM' quay wall existed of a rather high relieving floor in combination with a foundation consisting of a sheet pile wall and concrete piles. The concrete piles were placed at an inclination of 3 to 1. Some of them, the tension piles, were installed just above a clay layer whereas others, the compression piles, reached the underlying sand layer. A rather dense pile field was required to take care of the horizontal loads. An overloading of the tension and compression piles caused the observed excessive displacements. This overloading was either the result of settlement of the clay layer, due to the high surface loads, or the result of the relatively high horizontal earth pressure on the piles, due to high surface loads and the dense pile field.

The main characteristics of the current design are the relatively deep positioning of the concrete structure, which serves as a relieving floor, and the foundation of this relieving floor which consists of a combi-wall, concrete compression piles and steel tension piles (M.V. piles). The combi-wall, which also has a retaining function, is placed under an angle of inclination in order to reduce the tension forces in the M.V. piles and the earth pressure on the wall. Since the retaining heights have been increasing the use of stiff walls such as combi-walls became necessary. Adjusting the installation procedure to the local soil conditions optimized the derivability of the combi-walls.

The M.V. piles (tension) are placed at the waterside of the superstructure under an angle of 45°. This reduces the required number of concrete piles. M.V. piles are chosen because of their high tension load capacity, their rigid behaviour and their insensibility for corrosion. The driveability of the original M.V. piles is improved by reducing the tip cross section of the piles and by using larger steel cross sections than necessary for tensile loads.

An extensive test program made it clear that the used concrete (prefabricated and prestressed) piles developed a bearing capacity, which is substantially higher than the bearing capacity according to the Dutch Code. The local authorities agreed in using the higher values for design-purposes the concrete piles were applied as such in the design of the quay wall.

The current design of the quay walls appeared to be technically satisfying as well as cost effective.

The last presentation, partially given by Prof. A. Gens (Spain) and partially by Prof. E.E. Alonso (Spain), professors in Geotechnical Engineering at the Polytechnical University of Catalunya in Spain, dealt with the geotechnical design and construction of breakwaters in the Bilbao Harbour. Bilbao Harbour, which is located on the northern coast of Spain, needed to be expanded in the early 90's as total capacity was reached. The plans to build the required new harbour facilities consisted, among other things, of a 4.35-km long breakwater, which was designed for the protection of the harbour against the prevailing northern storms.

In order to ensure a proper design an extensive program of geotechnical survey, consisting of site investigation and laboratory tests, was carried out. This site investigation was set up in two stages. Initially a seismic reflection technique was used to gather the stratigraphical structure of the ground. Subsequently boreholes and piezocones were carried out to obtain more detailed information of the subsoil. The soil profile mainly consisted of sandy layers. Only a few clayey layers were detected of which the upper one, a clayey silt layer, turned out to be the critical layer in assessing the stability of the breakwater. Rather a large number of laboratory tests have been performed on this layer. For the design of the breakwater the tests concerning the (shear) strength of the upper clayey silt layer were of great importance.

The analysis of the static stability made clear that the safety factor for the leeward side was rather low, unless a replacement of the clayey silt layer by granular material was chosen. This solution however couldn't be adopted for environmental protection reasons. Therefore further study of the stability of the breakwater was required. First of all the wave impact forces have been analyzed, using scale model tests, in order to get a more precise force time action on the breakwater for the design storm. A static stability analysis using the results of the wave impact forces study showed that no equilibrium could be found except for high values of c_u/σ'_v. The positive effects of a berm of normal proportions, which was added adjacent to the leeward side, were limited. In order to solve the problem of the stability of the breakwater a different approach has been adopted. A dynamic analysis of the breakwater was carried out using the permanent deformation of the breakwater as the failure criterion and using impact forces related to the significant wave heights of a design storm. This analysis showed that for regular values of c_u/σ'_v relatively small toe berms were sufficient to achieve stability of the breakwater. In addition to the dynamic analysis attention was paid to the risk of liquefaction. The main results of a finite element stress analysis, using the cyclic shear test results as input, were plotted in a so-called cyclic interaction diagram. There appeared to be no risk of liquefaction when a berm on the lee-side was applied.

Discussion Session 2.1 – Harbour Works and Bridges

Chairman: Dr. T. Orr (Ireland)
Discussion leader: Prof. J.S. Steenfelt (Denmark)
Secretary: G. Hannink (Netherlands)

In the opening Dr. T. Orr mentioned that for this session 16 papers were submitted, 5 on bridges, 5 on harbour works, and 6 on various other subjects. 13 papers can be marked as case studies. Prof. J. Steenfelt posed a number of questions to the audience as guidance for the discussion:

- How can we introduce control and monitoring?
- How can we isolate site specific from general knowledge?
- What is the yardstick for different solutions in design?
- What is the yardstick for numerical verification tools?
- What is the yardstick for soil investigation methods?
- What is the benefit from GIS in planning and execution?
- How do we employ and benefit from geophysics?
- How do we transfer knowledge from other fields to obtain innovative solutions?
- What is the impact of environmental considerations?
- How does QA influence projects?
- Will EUROCODES change our daily life with regard to plastic design, increase in laboratory work, or assessment of characteristic values?

Steenfelt presenting a case study on the Stignaes Coal Terminal in Denmark with special focus on the instrumentation and monitoring opened the discussion. The Coal Terminal was completed in 1979 and extended in 1982, facilitating ships with a depth of 18 m (compared with 16 m in 1979), and a tonnage of 180,000 ton (compared with 125,000 ton in 1979). The design for the extension comprised a second retaining wall in front of the existing wall constructed from the bottom of the harbour. The quay wall was instrumented because of the innovative solution, because failure of the quay wall would be disastrous, and quay walls must remain operational during construction. Monitoring for this project, and in general, is important because:
- It serves as an early warning system;
- Predictions for the deformation behaviour can be made in a short period;
- Data can be used as bench marks for calculations;
- Data can be used for future designs and modifications.

In fact, monitoring forms the base of the observational method. A test site with a length of 25 m was chosen along the quay walls. Piezometers were installed in front of and at the back of the quay wall. An inclinometer was welded to a sheetpile. An alarm controlled the deformations; an audio-visual data-acquisition system was connected to the measuring devices. The anchors of the retaining wall were tested and it appeared that they were close to failure. However, rather than the anchors failure, the test site was disturbed by stupidity in connection with the cranes. Prof. Steenfelt concluded that Control and Monitoring are essential and must be introduced at the design stage. Control, however, has the image of being expensive and is therefore not popular; it needs an attitude of interplay. Instrumentation can be very valuable and can be cost effectively done.

In his presentation Prof. M. Boulon (France) pointed to some aspects of present day Geotechnics:
- Location of project-sites: more and more projects are located near unstable slopes, or on loose sediments or in regions with seismic activities;
- Spirit of present projects: nowadays sites are accepted that were declared not suitable for construction in the past;
- Soil and site characterization: geophysics is becoming more and more important;
- Soil improvement versus mode of failure: a suitable strength hierarchy should be established between the components of the structural system; the structural system must be able to mobilize energy;
- Physical modeling: comparisons must be made between the prediction (method) and the measured data.

Prof. Boulon illustrated his contribution by describing the results of a 1g-model test consisting of a landslide of limited thickness that meets an obstacle.

Dr. J. Clausen (Denmark) presented his paper 'Combined offshore investigation methods' which is printed in Volume I, pages 521-525, of the Proceedings. He described a new submersible soil sampling rig that is capable of combining a traditional boring system (three 6 m sampling tubes) and the vibrocore system. M. Maurenbrecher (Netherlands) posed the question to Dr. Clausen as to whether geophysical investigation methods were added to the rig. Dr. Clausen answered that this was being considered, but, as yet, had not been implemented.

Dr. E. Bellendir asked how it is established that samples taken by the rig are undisturbed?
Clausen stated that full recovery is checked, and so far it has been almost 100%, and the samples are visually inspected when they are being described. The samples appeared to be suitable for triaxial and oedometer testing.

B. Obladen (Netherlands) reacted to the call by Prof. Steenfelt for the promotion of monitoring. He stated that consultants and contractors will consider instrumentation and monitoring seriously, but that, because an increasing number of situations, where problems have occurred, are being settled out of court by insurance companies, the opportunity to learn by the mistakes of others, which is preferable, is offered less frequently. This tendency seems to exist nowadays in all Europe.

The contribution by Prof. Scarpelli (Italy) was on 'Slope movements and their effect on serviceability' with a focus on Monitoring Results and Numerical Modeling. It concerned slopes in overconsolidated clayey soils in which inclinometers were placed, and the relation between slope movement and rainfall and pore water pressure. Based on the results of his calculation model and the monitored data, he concluded that slope movements are linked with rainwater infiltration, prediction of this phenomenon is possible if the physical behaviour is adequately represented. Deep drainage as a mitigating measure in this case is detrimental.

In his contribution Dr. Bellendir (Russia) presented the plans for a Flood Protection Barrier near St. Petersburg aimed at preventing the flooding of an extensive area. The barrier will be 25 km long and will comprise 7 bridges, 11 dams (varying from 1200 to 3000 m length, with a crest height of 6.6 m and a crest width of 29 m), 6 water passes (each 240 to 290 m long), 1 tunnel (3,5 km long), and a breakwater 8.5 m high. Two ship passes are planned with widths of 200 and 110 m, and water depths of 16 and 7 m. The problems associated with the realization of this project are the different regulations in Russia and the rest of Europe, and the difficulty in obtaining the necessary budget allocation. This project will take 10 to 15 years to complete; stage 1 is expected to take 3 to 4 years.

Discussion Session 2.2 – Static and dynamic soil-water-structure interaction

Chairwoman: Dr. S. Lacasse (Norway)
Discussion leader: Dr. R. Jardine (United Kingdom)
Secretary: Dr. P. van den Berg (Netherlands)

Dr. S. Lacasse opened the session with an introduction, Dr. R. Jardine gave an overview of the theme of discussion session and next, the panelists presented their contributions:
- Excess pore pressures beneath embankment by Prof. Z. Lechowicz (Poland);
- Bridge piers subjected to ship impact by Dr. O. Hededal (Denmark);
- Dynamic pore pressures under river embankment by Prof. F.B.J. Barends (Netherlands);
- Importance of stress-path testing in determination of cyclic shear test by Dr. K.H. Andersen (Norway).

Overview of theme of discussion session by Dr. R. Jardine. The total number of papers in this session is 34, which can be divided into the following sub-themes:
- Static interaction (16),
- Cyclic interaction/fluctuation (8), sub-divided into: Seasonal groundwater cycles (1); Free water/stored products (3); Urban groundwater (1); Marine cyclic loading (1); Constitutive behaviour (2),

- Dynamic interaction (10), sub-divided into: Vehicle/ship impact (1); Pile driving (2); Machine vibrations (1); Earthquake loading (2); Breaking waves (2).

Most papers in the session deal with the analysis of practical problems like piles, tunnels, retaining walls, dams, etc. The active topics at the moment with respect to soil-water-structure interaction, which are covered in the session related contributions are:
- More complex geometry's (3D);
- Coupled groundwater flow:
- Unsaturated soils;

Whereas the following topics are not covered:
- Characterization of pre-failure behaviour of geomaterials;
- Local yielding, anisotropy;
- Time effects; delayed collapse, creep, rate effects.

Excess pore pressures beneath embankment by Prof. Z. Lechowicz. The contribution deals with the evaluation of test results and calculated results for a test embankment at the Antoniny site. The soil condition consists of a soft organic peat layer on a gyttja layer. Using a 2-dimensional large strain consolidation model, it was concluded that the prediction underneath the crest of the embankment was in good agreement with the test results. Going away from the crest to the sides of the embankment, the difference between calculated and measured results increases. It was concluded that (1) partial drainage could have a significant role and (2) no complete excess pore pressure may take place after final load application. In a discussion with the audience it was concluded that also creep effects and change in permeability during and after loading might have significant influence.

Bridge piers subjected to ship impact by Dr. O. Hededal. Starting with the example of the ship impact accident which took place at the Great Belt West Bridge in 1993, an outline was given of the modern design criteria for ship impact, which can be summarized in two statements: bearing capacity should be sufficient, and: permanent displacement/rotations should be limited. An approach and methodology was presented. Results of 2D elasto-plastic finite element calculations were shown. The failure mechanism for piers at the west side was dominated by sliding and at the east side by rotation (the foundation of the piers at the east side is deeper than at the west side). It was concluded that the interaction between the structural and the geotechnical engineer is very important: what is the governing mechanism, which constitutive models are applicable and, based on that, choosing the right numerical model.

Dynamic pore pressures under river embankment by Prof. F.B.J. Barends. In this contribution first an historical review was given with respect to the Dutch Water defense System. On overview was presented with regard to the appearance of floods in the past millennium, the development of laws, regulations, design practice and rules. At the moment some new trends are visible:
- Probabilistic design,
- From a (local) dike-cross-section approach to a (regional) dike-ring approach (the complete polder is taken into account),
- From failure analysis to functional analysis.

An example is given that the safety of a dike can be less during decrease of water level than at the peak water level (delayed failure). At the end it was concluded that information about all aspects of dike safety is very scarce. It is, therefore, appropriate to join efforts in the collection of international experience on failures.

Importances of stress-path testing in the determination of cyclic shear test are presented by Dr. K.H. Andersen. At the start of the presentation it was stated that stress-path testing for the determination of cyclic shear tests is very important for both coastal and offshore structures. During the presentation the attention was focussed on:
- The different stress path underneath and next to a gravity structure,
- The pore pressure generation during cyclic loading in different tests (triaxial extension, triaxial compression and shear testing),
- The effect of cyclic stress condition,
- (non-)symmetrical loading and anisotropy.

It was concluded that the cyclic shear strength depends on the test and on the average shear stress.

The use of contour diagrams, relating the average shear stress to the cyclic shear component, is a very convenient way to present cyclic data.

Discussion Session 2.3 – Dredging and Transport by Pipelines

Chairman: Prof. E. Togrol (Turkey)
Discussion Leader: M. van den Broeck (Belgium)
Secretary: H.J. Kolk (Netherlands)

M. van den Broeck noted that the Conference includes papers on dredged materials and on pipelines. However, no papers are presented on transported dredged materials by pipelines, This concerns transport of materials with bulk densities which are currently in excess of 1.35 t/m^3. Hydraulic engineering is not capable of adequately predicting the behavior of this soil-water mass in pipelines.

Geotechnical engineering profession is challenged in assisting the dredging industry on this issue. Conversely, it is noted that the dredging industry is nowadays capable of assisting geo-environmentalists and geotechnical engineers by improved control of the quality and quantity of dredged material:
– Denser soil-water mixtures (i.e. material with a lower water content) can be transported/dumped,
– Layers with thickness as low as 20 to 30 cm can be dredged within 5 cm accuracy,
– Techniques are available to separate sand and mud during dredging,
– Underwater dumping can achieve relative densities of dumped material up to 50 to 65%.

Prof. D. Zindarcic (Croatia) gave a summary of his conference paper. Densification of dredged materials consists of three stages: sedimentation, consolidation and desiccation. At present experimental and numerical tools are available to predict densification in these stages. This was illustrated by experimental data.

G.H. Van Raalte (Netherlands) challenged the audience with the following questions/cautions:
– How many boreholes and/or CPTs are required for a dredging area?
– Unlike for foundation engineering, upper bound soil parameters are required to estimate dredging production.
– Dredgeability requires an estimate of in-situ soil parameters as well as those of remolded soil. With respect to the latter, geotechnical advice is sought on conditions that govern de-integration of clay balls formed during dredging.
– The dredging industry needs to be informed on the occurrence of in-situ anomalies in dredging areas (e.g. shallow gas, wood, tires, and cables). This is generally not addressed in geotechnical/geophysical investigation reports.

T.J. Kvalstad (Norway) dealt with buckling of pipelines. Significant upward deformation of buried pipelines due to thermal expansion is observed. This depends on the laid pipeline geometry and soil cover properties. Significant uncertainties on predicting upheaval buckling exist.

H.J. Luger (Netherlands) noted that 'engineering judgement' and 'experience' might lead to erroneous designs. This was illustrated by two examples:
– A rough pipeline will experience a larger drag force from a mudflow parallel to the pipeline in comparison to a smooth pipeline. However, the opposite can occur for mudflow approximately perpendicular to a pipeline.
– Heat loss by convection from a buried pipeline is generally negligible. However, it can be shown that this is not the case in highly permeable soils and/or for large diameter pipelines.

The presentations and discussions identified issues alien to conventional foundation engineering practice.

Soil investigations:
- Dredging contractors can claim that their Clients have provided inadequate geotechnical information. To avoid this, it was suggested that Clients should give the dredging contractors the responsibility to perform soil investigations prior to bidding on dredging contracts.
- There is no unique rule for the number of boreholes/CPTs required for a dredging survey. The optimum solution might consist of a preliminary integrated geophysical and geotechnical investigation followed by a detailed investigation based on the results of the former.
- Upheaval buckling of pipelines is highly dependent on soil properties of as installed backfill material. These data are currently lacking and should be gathered to improve pipeline design and installation methods.

Research issues:
- Reliable criteria related to the pumping of dense soil/water mixtures through pipelines should be developed.
- Further research is required to develop design criteria for predicting de-integration of clay balls during transport of dredged material.
- Available design methods are inadequate to predict the soil resistance against upheaval buckling. Numerical, analytical and experimental work is required to improve these.

Advise to practicing geotechnical engineers
- Experimental and numerical tools are available for predicting densification of dredged materials. These can assist in developing proper disposal management strategies.
- Engineering judgement and extrapolation of experiences should be applied cautiously.

Main Session 3 – Highways and Airports

Chairman: Prof. F. Schlosser (France)
Co-chairman: Prof. V.A. Ilyichev (Russia)
Secretary: M.T.J.H. Smits (Netherlands)

This session on Highway and Airports illustrated the importance of geotechnics in the development of transportation infrastructure throughout Europe.

After the opening of the session by Prof. F. Schlosser, Prof. J-P. Magnan (France) in his keynote lecture placed geotechnics related to highways and airports in geographic and historic perspective. Prof. Magnan gave a fascinating overview of the development of highways in Europe and of recent airport developments worldwide. Especially in middle and east Europe a vast amount of new highway projects is planned. The need for highway construction in Europe, combined with the progress in construction techniques, contract procedures, environmental and time constraints and site specific risks (e.g. seismic risks), calls for constant innovation in geotechnical engineering and in site investigation techniques. The subsequent presentations of special projects in Finland, Austria and the Netherlands proved that this innovation is indeed apparent throughout Europe.

O. Arkima (Finland) presented several geotechnical and contractual aspects of the upgrading of highway E4 in the south of Finland. The novel and well-planned contracting method, that included private funding, was an important factor in finishing this project well in time. Mr. Adam (Austria) presented geotechnical aspects of the Austrian-Hungarian highway E-4, with a focus on re-use of excavated material in the road foundation. The quality of mixed in place cement stabilization works was controlled using state-of-the-art measurement and interpretation techniques. As a result of this, weak spots could be detected and repaired immediately. Mr. Huizer (Netherlands) gave an overview of the construction of the Schiphol Airport railway tunnel. This overview illustrated the vast amount of constraints for large infrastructure projects in densely populated areas.

The second keynote lecture in this main session was given by Prof. G. Houlsby (United Kingdom) understanding the behaviour of unpaved roads on soft clay. In contrast with the other presentations in this session, secondary or tertiary roads were the subjects. Prof. Houlsby clarified the complex (dynamic) load distribution in a layered system of granular material overlying soft clay.

At the time Prof. Ilyichev closed the session, the audience had been given a varied impression of geotechnical aspects of highways and airports, both in the design and in the construction of such projects.

Discussion Session 3.1 – Soft Soils

Chairwoman: Prof. F. Çinicioglu (Turkey)
Discussion leader: Prof. M. Nussbaumer (Germany)
Session Secretary: H. Pachen (The Netherlands)

Prof. Çinicioglu opens the session and introduces the discussion leader and panellists by giving backgrounds of their professional education and engineering/research capabilities. Prof. Nussbaumer starts with a short summary of the 25 papers that were submitted for this session. The following topics were covered:
– Site reports 44%
– Theories 20%
– Laboratory testing 16%
– Field testing 16 %
– Others 4%

Time allowed Prof. Nussbaumer only to mention a few of these papers in more detail.

In the papers dealing with 'Site reports' the paper of Akai and Tanaka on the Kansai international airport Osaka reports on settlements of more than 5 m. These settlements occurred in a formation of 200 m thick Pleistocene clay. Vertical stresses exceeded 500 kPa by approximately 30 m of fill on 20 m water depth. An interesting 'Gap-method' for an embankment widening is reported by van Meurs. The 'Gap-method' is a two-stage fill operation, aiming at the reduction of damage to the existing road. Unfortunately the effect on reducing settlement was limited.

In the group of 'Theoretical papers' several topics are dealt with, like a closed form solution for consolidation analysis in two and three dimensions by Castelli, and some other themes like radial consolidation around piles by Dr.E. Imre.

In the group of 'Laboratory testing' improvements are discussed by Alikonis for a shear box apparatus to decrease sample disturbance. A precision triaxial apparatus using small samples with 15 mm. diameter is proposed by Shogaki, where Jardine et al. proposed a hollow cylinder apparatus for embankment stability problems.

In the papers dealing with 'Field tests', Magnan reports on 3 tests embankments which are under observation for 17 to 32 years, having settlements up to 1.2 m.

Prof. Nussbaumer raised a questionnaire to the audience:
– How would you position our knowledge of soft soils?
– Where do we still have a lot of uncertainties with respect to design and evaluation? Possibly in obtaining realistic field parameters by laboratory and field tests?

- Can the disturbance of in situ soil conditions by sampling and laboratory tests, as well as in field test procedures, be evaluated?

To overcome difficulties for the construction of embankments or cuts in soft clays Nussbaumer mentioned, besides the application of counter weights (berms), the following ground improvement techniques: Piles; Anchoring; Vertical drains; Lime-cement columns; Deep soil mixing.

Dr. E. Imre (Hungary) states that the 1 dimensional radial consolidation model, as presented in the paper, can be used for the evaluation of the dissipation test data of the piezocone and other rheological type cone penetrometer test data. Models available can be categorised into 3 groups: uncoupled models; coupled models that lead to the uncoupled storage equation; fully coupled models. She elaborated the consolidation proces for a cylindrical cone. The presented models related to the first two categories (Soderberg 1962, Randolph and Wroth 1979) – besides that they have the same analytical pore water pressure solution - entail no total stress decrease on the shaft of the cone. The third category, presented in the paper (Imre and Rózsa 1998), entails a total stress decrease. The water pressure solutions were fitted on some measured piezocone data, concerning five different filter positions. The total stress response of the fully coupled model seems to be underpredicted in the light of a measured data set (Baligh et al. 1985). This can be attributed to the fact that the radius of the rod may slightly be decreasing due to the stress release of the rod.

During the discussion Prof. F.B.J. Barends (Netherlands) mentioned that a fully coupled consolidation model was already available due to the work of E.H. De Leeuw (Netherlands) in the sixties. However it appears that the publication of De Leeuw is only available in Dutch.

Dr. E. Farrell's (Ireland) presentation dealt with the calculation of the consolidation coefficient c_v from field settlement measurements and with the difficulties in estimating this parameter from field and laboratory tests carried out prior to construction. The value of c_v backfigured from three separate case histories of fill construction on soft ground were compared with those interpreted from laboratory tests using the standard oedometer and the Rowe cell, as well as from piezocone dissipation tests. This exercise showed the current difficulties in determining reliable c_v values for use in design prior to construction.

H.Wahls (United States of America) reminded that the consolidation coefficient c_v will decrease when of the number of loading steps increases. In the standard oedometer test the load is often doubled and in comparison with the field loading the magnitude of each loading step is too excessive. However it is known that even the c_v - value of the standard oedometer test is too low in comparison with field data. He mentioned that in his experience the interpretation of Taylor is preferable to the Casagrande method. Dr. Farrell observational method is therefore very valuable. With a minimum of effort a reasonable assessment of time settlement behaviour can be obtained.

Prof W. van Impe (Belgium) stressed the effect of disturbance due to sampling and putting the sample in the apparatus.

D.Tonks (United Kingdom) considered that the use of the traditional Terzaghi consolidation equation based on pore water pressure dissipation introduces the assumption of linearity, which is incorrect, and that this limitation is overcome by expressing the equation in terms of strain. Dr. Farrell agreed that there was some merit in that approach, however the prediction of the pore water pressure rates is fundamental in the design of structures on soft ground. Prof. Barends added that both terms are present in the consolidation equation and therefore the choice for either the pore pressure or settlement is not an issue.

Dr.A.M. Gaberc (Slovenia) resumes some experience in embankment construction on soft soil in Slovenia. Especially the highway across the Ljubljana marshlands was used to elaborate the predicted and measured settlements. By using an adequate field-monitoring program (measuring excess pore pressures, horizontal displacements, vane shear strength) it was shown that the ground improvement techniques, such as vertical drains, gravel columns and preloading, were successfully applied. Accelerated consolidation resulted in settlements between 2 and 3 m. during construction, while in a period of 10 years after completion only moderate settlements of about 10 to 20 cm, due to creep effects, have been observed.

Comparison between predicted and observed behaviour proves the suitability of rather conventional computational procedures regarding settlement and bearing capacity prediction.

R.Termaat (Netherlands) tried to answer the question *When is creep starting?* and *How to include creep in the settlement calculations?* He stated that we can use Mesri's (1994) 'End-Of-Primary approach', or for instance the approach described in the conference papers by Alén. When the EOP-approach is used the creep settlement or secondary compression starts when primary compression stops. The precise moment is when the slopes of the primary and secondary settlements –time curves are equal. Termaat pointed out that he believes that any separation of primary and secondary settlement is artificial. Actually secondary compression is the time integration of the creep rate. This creep rate is a function of the effective stress and the density or void ratio of the soil, which means that the creep will start during primary compression. However a fully integration of primary and creep settlements give as yet no satisfactory results when a validation with field measurements is performed. During the consolidation process large differences occur in the magnitude of the settlement (see for instance the conference paper of Jansen et al.). One of the reasons could be the underestimation of the permeability in the field compared to the permeability determined in laboratory.

Discussion Session 3.2 – Maintenance and Durability.

Chairman: Dr. L. Furmonavicius (Lithuania)
Discussion Leader: Dr. A. Loizos (Greece)
Secretary: J.A. van Twillert (Netherlands)

Dr. Furmonavicius, in his opening address, discusses the increasing demand for attention to maintenance and durability (or sustainability). The 34 session papers show this interest by their number and content. Dr. A. Loizos emphasised that the performance of road structures depends on the properties of soil materials, and, as such, for maintenance and durability the attention should focus on these aspects. Essential to structures in the field of geotechnical engineering for transportation infrastructure are a well-stabilised foundation, a proper drainage system, and reliable non-destructive control methods. The benefits of a new foundation support system for road embankments on soft soils that makes use of rubble fill, membrane interlocking, and pile arching was outlined by Dr. M. Topolnicki (Poland), whereas Dr. H. Thurner (Sweden) stressed the homogeneity in construction as the most important aspect related to durability and maintenance. In the evaluation of the homogeneity modern technology is introduced, like for the Continuous Compaction Control System equipped with most modern electronics.

The following vivid discussion of the session can be summarised by the following statements: There is quite a large gap between the knowledge of road (pavement) designers and geotechnical engineers. This causes problems in finding the most optimal road design, which can lead to extremely difficult and expensive maintenance requirements. Emphasised is the need to add more specific geotechnical education in the training of our road engineers. Consistent testing and long term monitoring of roads is seen as an essential component for optimal maintenance and durability.

Discussion Session 3.3 – Ground Improvement

Chairman: Dr. M. Stocker (Germany)
Discussion Leader: Dr. B. Berggren (Sweden)
Secretary: J.R. Deutekom (Netherlands)

About 120 people attended the discussion session. The 25 session papers dealt with known ground improvement techniques.

34 Recapitulation of the Conference

In his opening speech Dr. M. Stocker pointed out the growing importance of ground improvement techniques since the beginning of the century due to the increasinge lack of sufficient qualified building ground. In literature over 500 titles are nowadays available on the subject of ground improvement. He denoted the importance of quality and safety aspects throughout the process and the growing importance of environmental aspects.

Prof. R. Katzenbach (Germany) presented new experimental results and site experiences on grouting techniques. He stressed the growing need for design and safety concepts on ground improvement, the need for careful soil investigation and the need for a mechanical model on the grouting process.

Dr. H. Rathmayer (Finland) presented the highlights in development of the Swedish Lime Column Method. Much has changed between the first design guide which appeared in 1977, when only lime columns were used and the latest design guide which appeared in 1999 and deals with lime and lime-cement columns.

The final speaker Dr. M. Dysli (Switzerland) presented the growing importance of environmental aspects in the process of ground improvement. The influence of many human activities on the environment are nowadays no longer acceptable especially when ground water quality is at stake. He denoted the importance of working together by all involved parties to protect the environment.

The final half-hour discussion took place on lime-cement columns and vertical drains. In the case of lime-cement columns special attention was given to the proportion of the two elements. Finally one of the participants presented a new method to calculate the smear zone effects around flat vertical drains. For his contribution the discussion leader awarded him.

Main Session 4 – High Speed Railways and Subways

Chairman: Prof. M. Jamiolkowski (Italy)
Co-chairman: Dr L. Maertens (Belgium)
Secretary: K.J. Bakker (Netherlands)

Prof. M. Jamiolkowski opened with an inspiring introduction on the importance of the topic of this main session. He emphasised the merit for a country's economy to have a good system of both road and rail infrastructure. In most major Cities all over the world large investments are done to develop an adequate system for rail infrastructure, often underground, to facilitate the public transport. Besides that in Europe large investments are being done to introduce a system of high-speed rail transport to give an alternative to flying over short distances. This development is triggered by the increased demand on air transport capabilities and the limitations to increase these without too much environmental problems.

Subsequently Dr. R Leca (France) delivered the State of the Art report on High speed Railways and Subways. Dr Leca focussed on the geotechnical aspects of tunnels and track foundation. He observes that tunnels and underground structures become an increasing part of the works related to the construction of road and rail infrastructure. He discussed the difficulties inherent to tunnelling in relation with knowledge of the underground. He emphasised the importance of techniques to develop a 3-D view of the structure of the underground in order to develop techniques that adequately support the ground when excavating underground. In addition he discussed several techniques to evaluate bore front stability. After that Dr Leca discussed methods to evaluate the stability of the subsoil for track design on the soil surface, and methods to improve the stability, e.g. ground improvement techniques.

Prof. W. Wittke (Germany), gave a keynote lecture on The Tunnels of the High Speed Railway Line from Cologne to Frankfurt, co-authored by K.D. Eschenburg. The realisation of the planned 204 km long high speed railway from Cologne-Rhine/Main, with 24 tunnels with a total length of about 40 km, puts a great demand on the geotechnical engineers to meet all the problems in the subsoil to construct on the diverse locations. The different geological situations where discussed. After that the feasible construction techniques including shotcrete lined excavation type of methods (NATM) were discussed. Prof. Wittke discussed the mechanical methods to model the subsoil in an adequate way to cope with the encountered problems. Three projects where discussed in more detail.

As one of the special projects lectures, the geotechnical aspects of the Jubilee Line Extensions where discussed by L.F. Linney (United Kingdom). Linney discussed the difficulties which were encountered in the further construction of construction works after the Heathrow collapse where a shotcrete Lined tunnel (NATM) was under construction. On some locations one had to change the foreseen construction type to bored tunnelling, giving major delays. Linney discussed the particular geology of the tracks on the South Bank of the Thames with the Woolwich and Reading beds and the Thanet sand, in contrast to most tracks which formerly where build on the North bank with London Clay. He discussed the particular problems which were encountered near the Westminster Station where a deep building pit in combination with bored tunnels passed very close to the Big Ben Clock tower. Besides that the experiences with compensation grouting as a successful means to limit surface settlements where discussed.

After the coffee break the main session was continued, Dr L Maertens chaired the session. In a presentation Prof. J. Maertens (Belgium) spoke about the development of the High Speed Railway in Belgium. After a short introduction about the way that the new High Speed Railway in Belgium fits into the total system of High Speed rail tracks in Europe, Prof. Maertens discussed the geological problems related to this specific track, and the site investigations being used: CPT's, bore-holes and piezometers. In order to improve the soil, lime columns have been applied. At Arbres a viaduct on piles was constructed, for which at some places pile lengths of up to 60 m were needed to guarantee pile-bearing capacity. In Antwerp a tunnel has to be constructed lengthwise underneath an existing rail track.

Subsequently Prof. M. Kavvadas (Greece), who discussed the 'Experiences from the Construction of the Athens Metro' continued the session with a special project lecture. Due to antiquities and other archaeological material in the track of the new metro development, major restrictions were imposed on the design and tunnel construction which led to diversions and change of solutions. The main geology existed of Athenian schist and limestone. Due to large variations in the soil parameters one encountered more than once loss of bore-front stability of the open face bore-shields. The system used to resolve the problems was ground pre-treatment, and change of type of boring machine. At one location a shaft had to be dug to release a machine which had encountered major face instability.

The last presentation in this session was about the Copenhagen metro. In a twin presentation by Dr. J.H. Gravgaard and Dr. P. Jackson (both from Denmark), both the intentions of the principal of the project, by Dr Gravgaard and the experiences of the contractor, by Dr Jackson were discussed. The intention is to build a major underground metro line with a length of 8 km under Copenhagen, with minimum environmental impact, and no damage to buildings. Extensive site investigations have been performed, and a survey of historical data in the archives was made. Additional investigations by the contractor revealed a major valley in the underground limestone surface, with a large impact on the construction work. Groundwater models and monitoring helped to meeting the demands on limited settlements. This was also important, because recharge of the pumped groundwater was put in operation, and its effectiveness had to be established.

Prof. Jamiolkowski closed the session.

Discussion Session 4.1 – Railway Foundations

Chairman: Prof. G. Greshik (Hungary)
Discussion Leader: Dr. J.A. Lord (UK)
Secretary: H.G. Stuit (Netherlands)

Prof. G. Greshik opened the session on railway foundations by introducing the discussion leader

 and panellists and gave the scope of the programme of the session on railway foundations.

The discussion leader Dr. J.A. Lord went briefly over the 31 submitted papers for the session. Many old tracks are being upgraded now and many new tracks are being built for high-speed trains. The contributions for this session concern mainly the geotechnical problems associated with the original existing railway structures and with the improvements of stability of existing embankments. The problem of ground born vibrations and transient effects induced by passing trains on railways on soft soil profiles are highlighted. The combination of high speed and high axle loading render the behaviour of railway track formation in what is considered to be a complex soil-structure interaction.

V. Cuellar (Spain) gave a presentation on the improvement of the transition zones for high-speed railroad lines. Design recommendations for such lines demand very little settlements. This is a particular true for buried structures underneath embankments and abutments fills behind bridges. The use of hydraulic fracturing technique may achieve selective and gradual improvements of the deformation characteristics of natural soils and fills. He mentioned hydraulic fracturing technique when creating grouted of cement based stable mixtures, by means of injection through sleeve pipes. This technique could provide a valuable mean to improving the behaviour of transition zones of embankment and abutment fills. It may avoid the need to ballast corrections and their negative secondary effects. The Laboratoria de Geotecnica CEDEX has been involved in this research dedicated to correlate the basic parameters of the grouting process to the mechanical parameters of the treated material. This correlation was presented. In a response M. Rudrum (United Kingdom) gave a presentation on recent experience in stabilising old embankments in Great Brittain.

R.F. Woldringh (Netherlands) gave an overview of the special investigation test site 'No-Recess' at Hoekse Waard, Netherlands. At this site five different embankments have been constructed in order to investigate their feasibility for new railroad foundation construction methods on the soft soils in the Netherlands. The soil stratification at the test site shows a nine meter soft top layer of peat and clay. The five types of embankments include the following:
- A conventional embankment, used as reference,
- Scandinavian type of lime-cement mixed stabilised soil,
- An embankment on stabilised soil walls,
- An embankment on geotextile coated sand columns,
- An embankment on stabilised soil on small piles.

During the presentation all the embankments were shown as well their way of construction. The embankments, all built in 1998, will be monitored until 2000. In response to several questions about the design considerations and follow up, Woldringh answered that, after a thorough inspection and elaboration results will be published and proper selection of embankment construction can be made for the high-speed railroad line in correspondence with the local soft soil condition.

C. Madshus (Sweden), who replaced P.E. Bengtson, showed the results of measurements performed in Sweden for the high-speed line. In 1997 serious vibration were observed on the West Coast Line near Gothenburg. This vibrations only occurred during passage of the Swedish high-speed train at certain spots. It led to a research programme that included measurement of vertical deflections, pore pressures, local acceleration and velocities at the Lesgard test site. As measured accelerations of the Swedish high speed train were significantly higher than those measured from cargo trains, it suggested that the phenomenon was due to speed and, in fact, the 'critical velocity' for Rayleigh wave propagation for the soil profile was attained. The aim of this monitoring programme is to understand the mechanisms of behaviour, to evaluate the soil degradation and to give input to better design recommendations. Some questions from the audience put attention to possible ways for measuring critical velocities. Still quite some research is needed to solve the problem of critical wave speed.

Discussion Session 4.2 – Design and Construction of Tunnels

Chairman: Prof. G. Barla (Italy)
Discussion Leader: Prof. R.J. Mair (UK)
Secretary: L.E.B. Saathof (Netherlands)

After welcoming the audience of about 150 people, Prof. Barla introduced the people on stage. Prof. Mair, who presented three main items to be discussed later, prepared the introduction to the theme of the session. He also mentioned that 37 papers were submitted for to this session. Pointing out that he was deliberately intending to provoke controversial discussion, he proposed the following discussion topics:
- Ground treatment techniques in open face tunnelling (e.g. soil nailing, jet grouting, pre-vaults) can be highly effective in controlling stability and ground movements, but can we predict these beneficial effects? Can we reliably predict the improvement in stability and the resulting ground movements?
- Well-controlled closed face tunnelling with TBM's (slurry shield or EPB) are capable of limiting volume losses (in sand one meets often values in the range of 0.5% and in clays in the range of 1-2 %) and ground movements to very small values (between zero and a few mm's). What range of volume losses should be allowed for in design? Can we relate these to ground properties, or do they depend on the machine characteristics and its operation as well? How do we allow for the 'learning curve' when higher volume losses can occur (Shirlaw 1996) presented a clear learning curve: in the first 100 m from 180 mm to less than 10 mm)?
- Is K_0 (coefficient of horizontal earth pressure at rest) of any relevance to the ground loading acting on tunnel linings? Recently in the Dutch Second Heinenoord Tunnel project the measured loading on lining segments was mainly determined by construction operation effects, much more than K0-values.

Next, Prof. Barla gave the floor to the four panellists.

Dr. C. Erichsen (Germany) presented a case history of a road tunnel of approximately 2000 metre length, which has been constructed in the city of Bonn-Bad Godesberg, Germany. The tunnel was driven according to the shotcrete method in a gravely sand with low overburden. The tunnel crosses several buildings and twice the German Railway Line ('Intercity-Line'). It was calculated that the stability was not assured using the standard execution method. In order to achieve stability a stabilisation of the heading was necessary. In finite element calculations one had to assume apparent cohesion to calculate a stable face. A ground improvement by slightly inclined horizontal jet grout columns in the face seemed a good solution, but the construction of it had also many disadvantages (vibrations etc.) so that it was abandoned. Erichsen stressed the importance of keeping the time between crown and invert excavation as short as possible. The tunnelling underneath the German Railway lines was carried out successfully without effect on train traffic.

Prof. P. Egger (Switzerland) treated also the stability of the tunnel face and wall and he stated that the stability might become critical when the strength of the ground is low with respect to the natural stresses. Ground inclusions have been used with success to improve the tunnelling conditions. For a long time, rock bolts have been installed around the tunnel wall. More recently, various types of inclusions such as glass fibre reinforced resin (GFR) bolts or jet piles, and canopies of pre-cut concrete elements or pipe roofs have been placed in or around the face respectively. But there still are questions about the optimisation of the support systems e.g. what are more beneficial: short anchors with close spacing or longer ones with bigger spacing?

Dr. L. Martak (Austria) stated that there are still many discrepancies between results of computation in the field of soil-mechanics and the real behaviour of soils. He told the audience he had learned from Borowicka that by adding a so-called structural condition to equilibrium and failure condition all inconsistencies between calculation and nature disappear in as far as the failure state is concerned. He illustrated the use of the Structural Condition to the equilibrium of open face tunnels, explaining the failure zones around the tunnel surface as logarithmic segments.

Dr. B. Rymsza (Poland) presented some work about the K_0 factor. In static calculations of a tunnel lining most people apply the earth pressure at rest. To determine lateral loading they assume the coefficient K_{0n} due to natural stress state in-situ. This assumption can lead to unsafe designing, especially in heavy over-consolidated soils. Paying attention on the hysteretic stress model of soil and on local stress-strain disturbances in ground during a process of tunnel construction, the coefficient $K_{0t} < K_{0nOC}$ is suggested to be taken in evaluation of bending moments and lining displacements.

After these presentations Prof. Mair asked the audience to give comments on the first selected discussion topic, that of ground treatment techniques in opens face tunnelling. The question is: Can we reliably predict the improvement in stability and the resulting ground movements?

R. Leca (France) commented on this topic that even if there were good models the important question is how to validate them. He drew attention to the question of how to cope with the aspects of construction method in the design-models. Dr N. Taylor (United Kingdom) supported the need for good data from projects which are very important to be able to validate models. He also mentioned that we could do centrifuge model tests in which (single) mechanisms can be investigated and analysed. Dr. G. Viggiani (France) said, give us data of cases then we can learn and calculate the real safety factor of constructions. Leca replied to Taylor that centrifuge testing itself is also modelling of the real world like numerical calculations. Prof. Barla mentioned the effect of construction methods like jet grouting, where due to the grouting activity extra deformations occur. Dr. Martak added to this that the most effective place for grouting is the bench of a tunnel. Dr. Erichsen mentioned that grouting sometimes could result in surface heave, but if well controlled it leads to zero displacement of the surface, even if there is low strength in the beginning.

Prof. Mair than asked to move to the next topic he introduced, namely that of volume losses and their relationship with machine performance and ground properties. Dr. M. Hamza (Egypt) mentioned a tunnel project in Egypt with a slurry type TBM of 9.5-m diameter, where the settlements were a function of the soil stiffness. He warned also for high grouting-pressures in TBM drives. Mair added to this that in soft clay soils grouting could result in excess pore pressures, and thus in significant consolidation settlements developing with time. In some cases, the resulting consolidation settlements could be many times greater than the initial settlements. K.J. Bakker (Netherlands) mentioned some measurement results of the Second Heinenoord Tunnel project. The volume loss was about 1 % (30 mm maximum settlement) and it seemed to be a function of the injected volume of grout in the TBM tail. In his opinion the details of the machine operation were more important than the soil properties.

Because of the time, Prof. Mair wanted than to make a switch to the last topic, namely the relevance of the K_0 (coefficient of horizontal earth pressure at rest) to the ground loading acting on tunnel linings. G. van Oosterhout (Netherlands) presented some results of the Second Heinenoord Tunnel: two rings were equipped with structural monitoring instrumentation. The measurements have shown that the assembly of the rings caused very significant initial forces and moments in the segments. Moreover, these initial forces and moments do not decrease significantly, as is often believed. Since conventional tunnel design models do not consider the effects of assembly on forces and moments, predicted maximum forces and moments in the lining appeared to be about half the measured values. E. Van Jaarsveld (Netherlands) stated that the measurements in the same project showed the settlements to be a function of the injected volume of grout and the loading on the lining to be a function of volume of grout, but not of K_0 or soil stiffness. Dr. Rymsza said that one could influence the stress state of the soil around a tunnel.

Because the discussion was very much alive in this session, it was continued well into the coffee break and, only because of the next session, Prof. Barla had to close this session. He thanked everybody who had made this session as lively as it was.

Main Session 5.0 – Heritage Lecture and Special Projects

Chairman: Prof. N. Krebs Ovesen (Denmark)
Co-chairman: Prof. P.A. Vermeer (Germany)
Secretary: Prof. F.B.J. Barends (Netherlands)

After a short introduction by Prof. Krebs Ovesen, he gave the floor to Prof. P.A. Vermeer who presented a concise memorandum on Keverling Buisman, the Dutch pioneer in the field of geotechnics. Prof. A. Verruijt was introduced and invited to present the Keverling Buisman Heritage Lecture.

Prof. A. Verruijt (Netherlands) introduced Keverling Buisman, his book and his creep formula, which - originally purely empirical - did not receive wide appreciation. Later, yet a fundamental approval of the creep formula was given (Mitchell). Keverling Buisman and later his successor Prof. G. de Josselin de Jong (Netherlands), both emphasised that for soil mechanics the best available mathematics should be applied. Verruijt's lecture is accordingly. He introduces the principle of pseudo time to express irreversible cyclic deformation and applies it to the problem of acoustic waves in soils using an analytical approach. His method is new, elegant and allows introducing histeretic damping characteristics in the dynamic behaviour of soil in a proper way. In contradiction to visco-elasticity the dynamic damping of soils show histeretic reversible damping due to the intrinsic friction in the granular matrix. The basic mechanism is therefore independent on the frequency, and this is the core of his approach. The effect of damping is shown on the propagation of dynamic (shear) waves induced by supersonic line loads (High Speed Train). Because of the damping of soil fast train tracks should be built on soft soil! His theory is being validated by field-tests in the project NO RECESS.

Dr. G.W. Plant (Hong Kong) was the resident engineer of Hong Kong Airport, having gained experience before in Southern Africa. He gives a wide outline of the immense 20 B$ project of Hong Kong Airport, which opened in 1998, focussing on the site reclamation which covers 12 km^2, the tunnels for entrance and on the project management. Because of the scale, a parallel tender table was used as benchmark for the final tender, for confidence in the contractor's proposals. The project started in 1993, but it took until 1997 before a proper road access was established. There were dredging activities and on land rock removal; trucks drove 12 MKm! There were stability risks at the reclamation faces on the soft marine deposits and massive concrete seawalls. Preservation of water was a demand that was complicated by the seasonal activities of the rivers. The land reclamation, a lump sum contract of 1.2 B$, required a rigorous project monitoring. It was completed in 2.5 year with success due to high level engineering and technical skill of government and contractor. The high production rates were the main concern. Many land subsidence markers, recording primary and secondary settlements of the underground and creep of all fills, noticed significant settlements. Total settlement amounts to 1 meter, residual settlements up to 0.40 meter. The government took full responsibility that prevented added risk money in the contracts. The site preparation is published in a special book as a tribute to all workers.

T.W. McNeilan (USA) gained experience in coastal and offshore projects in California, Los Angeles. He focussed on the San Francisco-Oakland Bay Bridge, the east span replacement, where the main difficulty lies in the earthquake requirement, i.e. to withstand a magnitude of 8, like happened during the Loma Prieta earthquake in 1989.

An integrated geotechnical approach including land and marine geophysical surveys, on and offshore drilling and in situ testing, was adopted which showed to provide efficient, cost effective, and accurate data for design and reconstruction. A local laboratory fastened the data incorporation in GIS. It shortened the site investigation period significantly, by 30%.

The graphical presentations of the data in 3D coloured plots were impressive and clear.

Poster session – the poster award

Honorary Senior Team (Netherlands):
W.J. Heijnen
K. Joustra
E.H. de Leeuw
Prof. A.F. van Weele

Authors of papers were requested to participate in a Poster Session, arranged in the Conference Hall, and permanently accessible to all participants during the duration of the conference. On Wednesday-afternoon of the 9th, the authors gave explications about their subjects and had discussions with colleagues, while no other functions were held. The second part of the afternoon was used to share drinks and snacks in the same area, so that discussions could go on, and this really was the case. The number of visiting participants was large as well as the number of discussions among them.

It was decided by the organisers to award prizes for the 3 most outstanding Posters and a small Committee of 4 elderly Dutch experts was appointed to screen all posters and to nominate the 3 winners. This Committee consisted of Wim Heijnen and Bert de Leeuw, former directors of Geo-Delft, Kees Joustra founder and former Managing Director of Fugro Engineers and Bram van Weele, former professor at Delft Technological University and Director of IFCO. The winners were announced and the prizes awarded during the Conference Banquet on Thursday evening by van Weele.

Almost 100 posters were presented, covering the entire subject of the conference. The experience has shown that a poster requires a very brief presentation, in order that a passer-by can understand the subject and its value for himself, while almost walking by and looking on for less than a minute. It is not advisable to present a blow-up of the paper, as time for reading is too scarce for that while competing with 100 other posters. Every author must be aware of the fact, that a participant to the Conference may spent 2 or 3 hours viewing posters and with 100 posters this gives an average of only 2 minutes per poster. During this very brief confrontation the participant of the conference should be able to decide whether the subject deserves more attention or not. So, to highlight its merits and to make people easily understand its subject, a poster presentation requires very thorough attention and special preparation from the author. The reward however consists of discussions with seriously interested colleagues, sometimes leading to a more frequent contact or even co-operation.

The Committee Members visited the posters on their own and each made his own judgement before the internal discussions started. It proved to be not so difficult to arrive at a common decision with regard to the winning posters. Prior to visiting, it was agreed that each poster would be judged from different angles: clearness of presentation, value of the subject for the profession and originality. On this basis each person should come up with a list of the 10 best posters. Winners were the authors H-J. Köhler, I. Feddersen and R. Schwab (Federal Waterways Engineering and Research Institute, Karlsruhe, Germany) with their contribution: *Unsaturated condition below the ground water table and its effect on pore-pressure, soil and structure deformation* which has been published on pages 1109-1115 (Volume 2). Their poster was nominated by each of the four Honorary Senior Tteam members. The first prize was a sum of 1.000 Euro.

The Committee was enthusiastic about the average quality of the posters, the number of posters, as well as the interest shown by the visiting participants to the conference. It is considered an important means of knowledge transfer, much easier accessible than the 3 volumes with papers, covering more than 2200 pages for the XIIth European Conference! The institution of a substantial prize for a winner will encourage especially the younger authors, which is considered to be of great importance.

Workshop 1 – R & D Policy, Planning and Funding in Europe

Chairman: Prof. J. Hartlén (Sweden)
Secretary: A.F. Jonker (Netherlands)

Prof. J. Hartlén opens the workshop with an overview about what has happened the last few years. There were some special meetings in several European countries. As a result of them, a proposal was made to start a European Thematic Network (EuroGeoTech) with financial support from the 4th Frame Work Programme of the EU. Unfortunately, the EC at Brussels decided that there would be no financial support for this Network. Nevertheless, there are many items to look for in the coming century, for instance:
– Development of the society requires new land for housing and infrastructure;
– Constructions in and on soft soil need methods for stabilisation etc;
– Recycling of material needs to developed;
– Cleaning of contaminated soil.

There is not only a lack of dissemination of knowledge between the European countries, but also a lack of research money. This makes it necessary to promote a co-ordinated approach of the geotechnical research and development within Europe. EuroGeoTech has to start in a proper way; it can help for instance for:
– Co-ordination of the ongoing geotechnical R & D;
– A better dissemination of knowledge;
– Doing research that is better focussed on the real needs from industry.

Prof. S. Steedman (United Kingdom) highlighted the Process Steering of Geotechnical R & D. Steedman showed the organisation of the ECCREDI, the European Council for Construction Research Development and Innovation and especially the TRA EFCT Network (Targeted Research Action on Environmental Friendly Construction Technologies). This TRA consist of 10 clusters; one of this Clusters is New Technologies in Geotechnical Engineering (Cluster 8). Current initiatives are:
– Geotechnical Network, Type 1 to link national research programmes and projects, focussed on brownfield sites and redevelopment;
– The Fifth Framework – requiring socio-economic benefits from geotechnical research;
– A construction Virtual Institute, providing information and advice to Clients, industry and academics.

In the lecture about 'Planning: Success and failure', Dr. S. Lacasse (Norway) pointed out the following factors to have success in R & D projects:
– Select projects which fit to the existing R & D Programmes of the EU;
– One person has the pivot function of the total project;
– Gamble on success rate of proposals;
– Industry involvement;

42 Recapitulation of the Conference

- Network of new contacts and clusters;
- Funding increases interest from the industry;
- Evaluate the results of the R & D;
- Collect the experience in doing the research together.

Points of special attention:
- It can be difficult to obtain industrial sponsors;
- Make in any cases an evaluation of the results;
- Prof. Schlosser asked for attention for the needs from Industry, for instance:
- New models for design;
- Experimental research co-operation;
- Determination of characteristic soil properties;
- Needs for research to use unsaturated soil (preferably a code of practice).

G. Dubbeld (Netherlands) focussed on the point how to be successful in getting financial support from the EU Programmes from Brussels. To be successful, it is necessary to show that results
- Are used by industry
- Will lead to significant technical progress.

Dubbeld showed the main points for writing proposals within the 5th Frame Work Programme.

In the discussion, attention was asked to start a European Advisory Geotechnical Committee to help for instance:
- For a better co-ordination of the ongoing R & D projects;
- The industry for formulating the right questions to be solved (what is necessary over 5 - 10 years?);
- For making proposals for R & D within the 5th Frame Work Programme;
- To promote geotechnical engineering to the European Politicians and focus on future R&D programme's of the EU;
- To be involved in the preparation of the 6th Frame Work Programme of the EU.

To be successful, it will be recommended that this Advisory Committee consist of representatives of the national societies.

Workshop 2 – Stabilisation of Landslides

Chairman: Prof. E.E. Alonso (Spain)
Secretary: J. Lindenberg (Netherlands)

The chairman Prof. Alonso opened the workshop and introduced the following programme:
- Introduction and progress of ETC1 'Stabilisation of Landslides' by Prof. E. Alonso (Spain);
- Contribution 'Prestressed anchors for landslides stabilisation and retaining structures' by Prof. H. Brandl (Austria);
- Contribution 'Behaviour and examples of soil nailing for stabilisation of slopes' by Dr. A. Guilloux (France);
- Contribution 'Combined grouting and nailing techniques to stabilise Carmona Parador by Dr. V. Cuellar (Spain);
- Discussion;
- Prof. C. Viggiani (Italy) who was invited to give a short presentation on 'Analysis of trench drains', was not able to attend the conference.

Rodriquez Ortiz and E. E. Alonso will complete the final report *Stabilisation of Landslides* of the ETC1 within a few months. The report is meant as a practical support for the designer. It contains relative simple methods for stabilisation including keys to select the most convenient method and design charts and examples. Prof. Alonso shows some pictures to give an impression of the report.

Prof. H. Brandl (Austria) thanks the chairman and members of ETC 1 for their efforts and memorises that this committee is on of the most active technical committees in Europe. Prof. Brandl starts his presentation with the remark that absolute safety, which is often asked or expected by lawyers and politicians, can not be realised by the technicians. Therefore he stresses the importance of risk analysis and risk assessment, explanation of the assumptions (most probable conditions) and of the policy including the background of stability or safety factors. He recommends the use of the observational method including monitoring which should be started before the actual construction. The observational method can be useful to 'prove' stability and may save costs in avoiding 'overdesign' as frequently present in conventional design. Brandl mentions that often 80% of the total costs are (invisible) in the ground. He mentions a large number of mechanisms, aspects, potential solutions and measures including practical examples. Among other things:
– Corrosion and required double protection,
– Weathered rock may be dangerous,
– Importance of reliable estimate of residual shear strength,
– Anticipation to possible remedial measures in future (observation and monitoring),
– Importance of drainage.
Illustrative examples of wall that may need additional anchors in future (tubes already present in case monitoring demonstrates creep), wall quickly constructed to release earth pressure after observation of creep near railway, very high bridge piers (up to 160m, very sensitive to deformation of foundation)

Dr. A. Guilloux (France) presents calculation methods for soil nailing based on the French experience. He distinguishes three types of soil nailing (micro pile, middle and heavy soil nailing) and shows some examples. His main conclusions are: there are many possible solutions. Important is the aim of the measures: To stop or just delay the slope movement; stabilisation for site with many data and frequent monitoring: $\Delta F/F = 5$ to 10%. For sites with less information $\Delta F/F$ should be 30 to 50%. For complete stopping of movement soil nailing is usually not sufficient; combination with drainage or other measures is required.

Dr. V. Cuellar (Spain) presents an interesting case of combined grouting and nailing for slope stabilisation. It concerns the Carmona Parador (middle age castle) on a sort of platform on top of a steep sloped rock consisting of calcareous sandstone. The lower part (with part of the castle) is slowly moving downwards. Previous stabilisation attempts in 1981 and 1987 were not successful because movements were still present afterwards. The visible cracks and the possible slide planes (including water entering the cracks) are presented and the measures that have been taken explained. These measures consist of soil anchors and fracture injection by grouting via tubes up to 50m depth. The grouting results in increase of effective stress and increase of soil/rock stiffness. After the measures were carried out the movements stopped and the Parador has been opened again.

Discussion: Prof. S. Cavounidis (Greece) asks if a relation could be established between the (increase of) stiffness and the (decrease of) slope movement. Dr. Cuellar answers that such a relation was not found. Prof. Cavounidis emphasises the importance of optimisation of costs, safety, lifetime and relating risks. He mentions that over 50% of anchor failure occurs in the anchor head (damage due to corrosion, water intrusion) and recommends being very accurate in isolating this part. Behind the anchor the cement often prevents head corrosion.

Workshop 3 – Eurocodes

Chairman: Prof. U. Smoltczyk (Germany)
Secretary: W.J. Heijnen (Netherlands)

The following papers regarding this topic have been sent in:

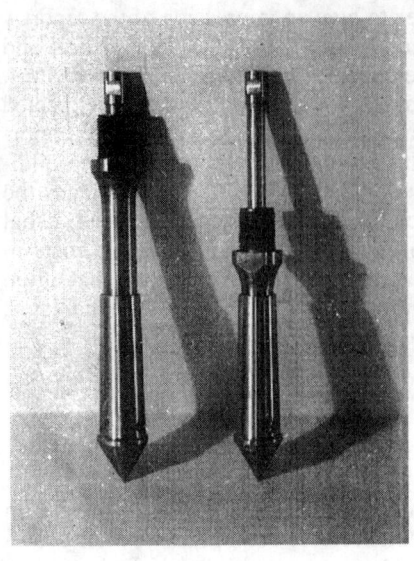

- A few thoughts about ultimate limit state verification following Eurocode 7, R. Frank & J.P. Magnan (France)
- Application of Eurocode 7, part 1 in the design of a piled foundation, W. J. Heijnen (Netherlands)
- The relation between Eurocode 8 and Eurocode 7, Pedro S. Sêco e Pinto (Portugal).

The workshop started with short introductions by R. Frank (Germany), W.J. Heijnen (Netherlands), B. Simpson (United Kingdom) and Prof. S. Sêco e Pinto.(Portugal). R. Frank focused on some of the principles in connection with the application of the design cases A, B, C and D (1 and 2). He explained the reasons for adding the cases D to the cases A, B and C as given in Eurocode 1. The main reason is to respond on the rigid wishes of Germany and France to get more freedom. W.J. Heijnen showed on the basis of a practical example the differences between the results of the application of the cases B, C, D1 and D2 to the design of a piles foundation. He also compared the results with the practice in The Netherlands. B. Simpson gave his view with respect to the factoring at the action side. He underlined his proposition that factoring the various action effects is a much more logical approach than factoring the sum of the action effects. Prof. Sêco e Pinto dealt with problems concerning the use of Eurocode 7 and Eurocode 8 on Earthquakes. He stresses his opinion that the compatibility between both Eurocodes should be improved in the near future.

Several participants took part in the discussions on various subjects regarding Eurocode 7, part 1. Because of lack of time discussion were very limited. Only a few problems could be touched.

Prof. N. Krebs Ovesen (Denmark) expressed his satisfaction that we now see that the important items of the Eurocodes are entering the discussions in conferences like this one. A very prosperous development indeed. In the past discussions on the matter were restricted to meetings of invited experts of the various countries of the European Community.

The workshop was closed by Prof. U. Smoltczyk as follows:

'Although I am sure that we are just on the way to get to core problems during our discussion, we now have to finish it in accordance with the schedule of the conference. This gives me the pleasure to thank all of you who took part in this workshop - especially rewarding our panellists for their most informative papers and also those who contributed to the debate'.

During the discussion on EC7 static problems we focused on the changes in subclause 2,4 as intended for the conversed EC 7, part 1 document. Basically the earlier method of verification, called the B+C approach, will remain valid in accordance with ENV 1997-1. But the approaches with only one verification flow, called C2 and C3 (formerly D1 and D2) will also be allowed. This was deemed to be necessary because the users of the code expect that sizing foundation elements should not become more expensive or less safe then they are accustomed to by long time experience. So we will have two alternatives for finding the ultimate resistance: either to put the partial safety on the input to computation models or on the output in terms of resistant forces. Since we have ultimate limit states in geotechnics which depend linearly on the shear strength (sliding, slope stability) and others based on non-linear models (bearing capacity, earth pressure) it cannot be expected that these different approaches will result in equal dimensions. It will therefore remain the duty of the national standardisation institutes to decide which approach is most compatible to national experience. They will also have to assess the partial safety factors on the basis of the suggested values in the Annex A of Eurocode 7, part 1. Harmonisation in geotechnical engineering therefore means, at least for the period foreseen for the trial application, harmonisation of the principles and the general ways of approaching the geotechnical problems.

As to the dynamic design in earthquake area, participants seemed not to feel sufficiently competent to discuss the critical remarks which were put forward by Professor S. Sêco e Pinto. It should be kept in mind however that Eurocode 0, Basis of Design, distinguishes between ULS by accidents and ULS by earthquakes. This implies that the assessment of specific partial safety factors in earthquake design situations does not contradict the rules of EC7. Consequently all aspects connected to failure caused by earthquakes have been intentionally left to Eurocode 8. Problems which have been put forward by our panellist, Professor Sêco e Pinto, will have to be treated in the working groups of EC8. Subcommittee SC7 will be kept informed. The exchange of information should become more current and intensive.

In conclusion, we all hope that this workshop has contributed to the willingness of geotechnical engineers in Europe to apply EC7, part 1 in their design work on foundations and other geotechnical structures. It is the only way to come to a set of consistent rules, which correspond within acceptable limits to the prevailing expertise in all European countries. I thank you for your attendance.'

Workshop 4 – Information Technology in the Geotechnical Profession

Chairman: Dr. B. Rydell (Sweden)
Secretary: L. de Quelerij (Netherlands)

This document provides a summary of the Workshop 4 Information Technology in the Geotechnical Profession.

Dr. K. R. Massarch (Sweden) gave an overview of the application of IT in the field geotechnical engineering. He distinguished three areas of development:
– IT-tools: hardware as driving force beats the software; the development shows the change from slide rule – computer - main frame - calculator- PC- database etc.
– Communication Information Exchange: development of communication speed from: Runner – horse – tram – car – air plane – telegraph – telephone – fax – electronic communication – data base; Future development will allow for multiple services on 1 line video/data/telephone communication and the wide use of database and knowledge based technology;
– System modelling: development from 1D-construction drawings (already from Egyptian time) – 2D perspective drawings – 3D-computer pictures towards computer modelling: there are no roads: roads are made by walking.

In addition the following speakers presented examples of 4 types of IT-applications dealing with respectively[1]:
– Information Retrieval and Communication by Dr. B. Rydell
– Validation of Computer Models/ Programmes for Geotechnical Design by Dr. A. Bond
– Geotechnical Design Decision System by M. van Veghel
– Education and Vocational Training by Dr. D. Toll

Dr. B. Rydell (Sweden) explained the three components of Information, Retrieval and Communication. Information is defined in this respect as geotechnical information needs for scientists, clients, customers and authorities. Retrieval takes place from libraries (including databases, web site searches), electronic journals, scientific societies, conference lists and from websites. Communication will increase by virtual conferences, e-mail, mailing lists and news groups.

The warnings of IT have been addressed by Dr Andrew Bond (United Kingdom) regarding the need for safeguards to ensure proper use of computer generated information. His concern refers to users without enough knowledge, communication gap between engineers and software designers, use of programs beyond their application area, insufficiently calculating process, unclear programme limitations. Dr. Bond suggests focusing more on the use of ISO 9126 in order to reduce

---

[1] These contributions are published in this volume, separately

risks. Quality assurance should therefore deal with software (quality standards distributed in 6 classes), hardware (especially problems related to data transfer and data storage), people (needed skills depending on their various roles) and the workflow process (hierarchy of modelling).

The modelling can be defined at different levels: Engineering models – conceptual models – computational models. The importance of acceptance criteria for these models is emphasised.

M. van Veghel (Netherlands) demonstrated the use of agents in a decision support system (DSS) for foundation design of buildings. The presented ongoing research, DESSYS fits into the encapsulating VR-DIS (virtual reality design information system) research at the Eindhoven University of Technology. In VR-DIS, a multidisciplinary design system is developed using agent- and VR-technology. By means of this multi-agent system different participants that are working on the design of a certain building can communicate and co-operate, supported by their own software-agent. The scope of the DESSYS research is to develop the agent which represents/supports the geotechnical engineer. This way the knowledge of the geotechnical engineer is represented in the early stages of the design-process and not only after the actual design is finished. Probably the most interesting aspect of the concept is deriving foundation-solutions out of a number of design-criteria given by the physical situation and the architect. Van Veghel had the opinion that this kind of knowledge based tools will have a major impact on the future design approach for buildings.

Dr. D. Toll (United Kingdom) gave a demonstration on recent educational and vocational training tools that are being used in geotechnical education programmes at universities worldwide. These tools consist of reference materials, animations and interactive demonstrations, simulations, games and knowledge based systems. Very interesting examples were shown on computer aided (CA) geotechnical learning materials from the University of Durham and the South Bank University of London. The last university has a very attractive demonstration on the performance of Proctor tests and a very humorous site investigation game including all types of practical communication challenges. Dr. Toll concluded that there are already good CA learning methods for geotechnics available to be used compliant to other teaching methods.

After the presentations Dr. Bond chaired the workshop discussion. Regarding the topic Information Retrieval and Communication question marks were placed at the quality assurance of data and information (how does a user know the reliability of the data) distributed from the web). Also the question related to the terms of payment of data ware housing and data mining was addressed. The aspect of information overflow and the use of intelligent search options were discussed including the positive effect of serenpendicity.

Warnings were given by the audience not to give away all your information because this information forms a major part of the company capital.

Regarding the item Validation the need for delivery guarantees from the software supplier was emphasised. Also the use of benchmarks developed by the geotechnical society (ISSMGE benchmarks) were discussed although primarily the software should meet the specific requirements from the client users.

From the discussion on the use of DSS systems it can concluded that a strong development in the use of this type of knowledge based systems can be expected. Remarks were made on the potential of including local experience of senior experts in an easy way. There should always be added value from senior experts. At the end of the workshop the speakers agreed upon the intention to form a group that will consist of people that are doing research on or are interested in 'IT in the geotechnical profession'. More information on the presentations is summarised on the SGI website (swedgeo.nl).

Workshop 5 – Geotechnical Engineering Education

Chairman: Prof. I. Manoliu (Romania),
Secretary: Prof. A. Verruijt (Netherlands)

During this session the state of geotechnical engineering in Europe was discussed. Presentations were given by Prof. J.P. Magnan (France), Prof. R. Katzenbach (Germany), Prof. A. Verruijt (Netherlands) and Prof. I. Manoliu (Romania). The speakers presented the developments in geotechnical education in their respective countries, from which it became clear that there are some general trends, such as the introduction of environmental geotechnics, and the use of modern teaching methods in the curriculum. Prof. Manoliu also presented the results of a study of the curriculum and the contents of geotechnical education in European universities, both on the undergraduate and the graduate level. Despite many local differences it appears that the scope and the contents of geotechnical engineering in the various European countries have very much in common.

Workshop 6 – ERTC-3 'PILES'

Chairman: F. de Cock (Belgium)
Secretary: H.J. Luger (Netherlands)

The chairman, F. de Cock, opens the session, welcomes the audience and explains about the past and future activities of the European Technical Committee 3 'Piles'. He highlights, in particular, some of the aspects which the committee faced while performing its task:
– The large gap between science and practice;
– The changing regulations as are now being laid down in Eurocode 7 and the National Application Documents (NAD's) in the various countries;
– The disappearing borders between countries.

De Cock subsequently explained briefly the contents of the reports produced by ERTC-3 so far. The latest product of the committee, entitled 'Survey report on the present-day design methods for axially loaded piles. European practice' was published at the occasion of this XIIth ECSMGE and could be distributed to all participants thanks to the support of the Belgian Society.

One of the problems which is encountered is the large variation of definitions and types of limit loads. Within the next 2 to 3 years the ERTC-3 aims to produce a report with recommendations for execution and interpretation of static load tests. Goals which are aimed for in that report are:
– Conformity with Eurocode 7;
– Clarification and harmonization of definitions;
– Introduction of quality levels, which must be chosen by a qualified engineer;
– Harmonization of test procedures, depending on quality level.

Dr. A. Mandolini (Italy) spoke on the relevance of the static load tests for the analysis and prediction of soil structure interaction. Often 1 to 2 percent of the piles is tested in a single foundation. When this is the case a series of pile load settlement curves is available as well as insight in the variability of these curves. This variability influences (or should influence) the design. On basis of a large number of load tests, as available for different pile types, the influence of foundation structure flexibility on its vulnerability to variations in the axial pile stiffness.

Prof. Smoltczyk (Germany) presents the essence of the design philosophy on basis of estimated or measured load-settlement curves. The load settlement curve must be interpreted to derive both design values in terms of settlements and in terms of maximum load (ULS or ultimate limit state). German practice is to focus more on the settlements than on limit loads and capacities. However, either of the two criteria can be governing. This implies that not just the maximum load should be registered, but that the full load-settlement curve is to be taken from the pile load test and to be analyzed.

The effect of pile load test specifications on the outcome of the test and its interpretation was discussed by Dr. M.G. England (United Kingdom). Items like the way the load increments are applied and subsequently the load are kept constant for a while influence the outcome of the pile load test. Already a 1% change in load may give a noticeable change in pile response. In particular the distribution between shaft and tip resistance leading to a total response is important. There is a change from primary to secondary response depending on load level and time during testing. Higher load application rates tend to increase the apparent capacity of the pile.

There was a question from the floor on the way Dr. England derived the creep load from his test procedure. The response was that this creep load represents just a 'point' on the load-settlement curve and is not crucial in the interpretation of the pile load test. Time prohibited the discussion on the item to be continued.

Workshop 7 – Pavements

Chairman: Prof. A. Gomes Correia (Portugal)
Secretary: F. de Boer (Netherlands)

This workshop dealing with geotechnical aspects in design and construction of pavements and rail track is closely related to ETC 11.

The objective of the workshop was to have an international overview of the more recent advances in modelling and testing pavement foundations and rail track and to enhance the exchange of knowledge between transportation and geotechnical researchers and practitioners in this field.

For 35 participants 5 lectures were presented, preceded by a short introduction of the chairman of ETC 11, Prof. A. Gomes Correia, who mentioned the terms of reference of ETC 11 and introduced the invited speakers.

Prof. S. Brown (United Kingdom) was giving an extensive overview of the contribution of soil mechanics to pavement and rail-track foundations. He addressed the following key points:
– Typical soil mechanics features characteristic to pavements:
  – Soil generally above the water table
  – Repeated applications of stress below failure (small number at high level during construction, large number at low level in service)
– Interest in resilient response to repeated loading and to accumulation of plastic (permanent) strains in soils and compacted granular material.
– Resilient response is known to be markedly non-linear, found from laboratory repeated load triaxial testing, from field-testing and from in-situ measurements.
– Linear elastic theory is generally used for pavement and rail track design but this has limitations. The non-linear properties of sub-base are influential to the asphalt layer of flexible pavements, the subgrade properties much less so.
– For back-analysis of surface deflections obtained with the Falling Weight Deflectometer in pavement evaluation procedures, it is very important to account for non-linear soil properties.
– Finite element analysis can be used to more accurately model pavement response to load accommodating non-linear response.
– Pavement research has focussed on resilient modulus as a function of applied stress, whereas the relationship between shear modulus and shear strain has been a characteristic of earthquake engineering, which also involves cyclic loading. More use could be made of findings in this field for pavement applications.
– Various concepts for research need further study:
  – Threshold stress (below which, plastic strain accumulation is negligible)
  – Rate of loading effects in clays
  – Use of new theoretical framework for partially saturated soils
  – Use of the Hollow Cylinder Apparatus for application of more realistic stress paths than are possible with the triaxial configuration.

Prof. Brown concluded that in addition to the above, developments in field testing and the application of improved theoretical modelling, particularly with regard to dynamic effects, should be proved.

Prof. A. Gomes Correia (Portugal) held a lecture about advanced testing and modelling of unbound granular materials, which could be seen as a summary of Lisbon workshop, 1999 January. The proceedings of this workshop *Unbound Granular Materials. Laboratory testing, in situ testing and modelling* is published by A. A. Balkema. He stressed the need to move away from empirical tests and methods to mechanistic approaches. This must be done at the construction and service stages. At the construction stage specify stiffness and water content (or suction) as end-products requirements for the complete foundation improves quality and ensure mechanical design requirements. The use of mechanical in situ tests allows analysis of test results to be done by a theory (modelling) and in consequence to derive parameters that can be used in design. Comparison of in situ tests and laboratory tests must be done for the same stress and strain levels. The main recent developments in laboratory testing show that:
- Local measurements of strains from less than 0.001% to several % can be done accurately.
- Cyclic loading or unloading-reloading cycles for different stress states are suitable to evaluate quasi-elastic material properties.
- Inherent and stress induced anisotropy are important.
- Stress state dependency of quasi-elastic stiffness is important.
- Isotropic stress paths and constant confining stress paths can be used to model the quasi-elastic behaviour observed under variable confining stress paths, more representative of in situ stress field.

Prof. Gomes Correia identified the needs for future research in two main areas: 1. Modelling permanent deformations for more than 10^6 cycles and related test method simulating the rotation of principal stresses in the field; 2. Internal stress-strain measurements in pavement layers to validate modelling and design.

Dr. A. Quibel (France) told about the need of simulation tools for automatization compaction process. He introduced the many advantages of the embarked devices in the compaction equipment. As an example he shows how GPS devices improved the compaction process and could identify zones where the specified number of passes has not been achieved. Dr. Quibel stressed the idea of developing a simulation program to help in selecting future technology and to achieve better compaction processes than available in current practice. He also pointed out the need of field trials to calibrate the simulation code. Finally he suggest that this work must be developed at a European level involving engineers, researchers, contractors and owners.

Dr. D. Adam (Austria) has lectured about continuous compaction control and modelling of the dynamic problem. He explain the basics of continuous compaction control (CCC) with vibrator rollers and how modelling the soil-interaction problems. This modelling, using the substructure method is used to determine the process of a vibrator roller compacting homogeneous or layered soil. Results are presented both in the time and in the frequency domain. Dynamic compaction values of roller integrated continuous compaction control systems are calculated and the process is shown according to the operating condition of the roller. Dr. Adam pointed out that the increase of the shear modulus causes a change of the condition influencing the dynamic compaction values significantly. A case study (Austrian-Hungarian highway A4) was presented to illustrate that CCC provided an improved quality and could detect weak spots, which could be repaired immediately.

Prof. W. Haegeman (Belgium) has given a lecture about the subject of non-destructive quality control of road and railway constructions by SASW and Georadar measurements. He pointed out that Spectral Analysis of Surface Waves (SASW) and the Georadar are non destructive techniques base on wave propagation which nowadays are used more and more for the quality control of road and railway constructions or the detection of subsurface obstacles. Both complementary techniques are used for definition of interfaces between different material layers and for categorising different material types. After a short introduction on the working principles of SASW and Georadar, case studies were presented to illustrate the possibilities of both tests. SASW can be used to assess the stiffness or elasticity modulus of the individual materials so a quality control of the compaction degree is easy to perform. Continuous radar profiles of roads allow for the evaluation of the condition of pavement structure, asphalt pavement quality, drainage or local damages. High-resolution radar images increase the capability of the system to detect thin layers and small objects like pipes or even reinforcements. Prof. Haegeman concluded saying that integrated analysis of SASW and Georadar measurements are the base for an advanced road and railway evaluation system.

After the five lectures by the members of ETC 11, discussion was open and two main contributions came from the floor.

Dr. Pauli Kolisoja (Finland) has contributed with a presentation related with the simulation of the mechanical behaviour of unbound granular materials by means of distinct element method (DEM). He explain that simulation tests made using DEM provide a useful tool on investigating the basic mechanical behaviour of granular materials used e.g. in unbound layers of road pavements. In comparison to normal laboratory tests the DEM approach has a number of unique features. For instance, as the shape and position of each individual particle as well as the forces that are transmitted between the interacting particles are all mathematically defined during the simulation, a DEM model enables e.g. the distribution of stresses, local changes in density and actual deformation patterns inside of a granular material to be studied. Further, a DEM model enables also the effect of intrinsic material properties of aggregates to be studied, because for instance both the stiffness and surface characteristics of the particles can be easily varied. On the other hand, however, it must be realised that even if DEM simulations can for the moment be made both in 2D and 3D using models that consists of several tens of thousands of particles or even more, they are still fairly far from being able to fully reproduce the complexity of the aggregates that are used in real pavement structures. Therefore it can be concluded that:
– DEM is a very useful tool in improving our understanding on how the granular materials are behaving and why they are behaving as they do;
– DEM is not replacing the normal laboratory testing methods but supplementing them;
– DEM simulations are fairly computing intensive; therefore simulation of cyclic loading conditions is somewhat time consuming.

Prof. S. Marchetti (Italy) presents the use of in situ dilatometer (DMT) testing to layer evaluation and some useful correlation with other parameters used in pavement engineering.

After a summary of the session by Prof. Brown, Prof. Gomes Correia closed the session with a call to the participants to make known their interest for geotechnical pavement knowledge and/or geotechnical rail track knowledge.

Special Workshop – Investigation Methods

Chairman: Prof. J.D. Nieuwenhuis (Netherlands)
Secretary: J. Puechen (Netherlands)

The Special Workshop Investigation Methods was held on Tuesday 8 June, 18:00 to 19:30 hours as a tribute to new advances in soil investigation methods.

T. Lunne (Norway) gave a presentation on the updated International Reference Test Procedure (IRTP) for the Cone Penetration Test (CPT). The IRTP is part of the proceedings of the XIIth ECSMGE (hard copy and CD-ROM) and will in future be available on the Internet. The historic development of the IRTP showed key dates of 1977, 1989 and 1995. The last date represents the start of the work of the now published version of the IRTP. Key features of the update include full integration of the piezocone test and implementation of accuracy classes for the results of the test. Recommended piezocone features include a filter position in the cylindrical extension of the cone, data processing to obtain derived parameters such as pore pressure ratio B_q, and details of piezocone dissipation testing. The selection of the accuracy classes relates primarily to the geotechnical use of the data for particular soil conditions. Class 1 will be difficult to achieve at present, but provides direction for future improvement efforts. Most of the current practice falls within Class 3 or Class 4. The IRTP document shows main text and notes. The main text includes 'shall' phrases and the notes provide clarification. The discussion session included further explanation of the philosophy: prescriptive requirements for equipment and procedures versus performance for end results. Concern was expressed about situations where companies claim to test to a better class than that actually achieved and about price differentiation per accuracy class. Guidance about demonstration of meeting the requirements of a particular accuracy class will be necessary in practice, together with 'educational' activities.

C. Bremmer (Netherlands) gave a summary of the recently introduced Geotechnical Exchange Format (GEF) for CPT results. The software and hard-copy guidance was made available free of charge to the workshop participants. The Dutch CUR research organisation supported the software development that arose from a need to achieve a practical and complete exchange platform for digital data. Important steps in the development included a questionnaire round and a review of already existing Dutch and other national digital formats. The principal tools covered by the GEF are a verification module, a dynamic link library, support of EXCEL spreadsheets and free software downloading from the Internet. Demonstration of the operation of the GEF included the procedure for syntactic verification, graphical data view and inspection of the data file. The GEF set-up allows extension to other geotechnical data, such as laboratory test results. The discussion session covered topics such a file management on the basis of time labels, availability of penetration rate data and compatibility of the format with the requirements of national standards for the cone penetration test. The strategy for free distribution of the software seeks a greater use of the improved format, including possible national and international standardisation.

E. Farell (Ireland) presented a summary of the ISSMGE recommendations for geotechnical laboratory testing, prepared by European Technical Committee ETC5 and recently made available through DIN publishing. The document is in three languages: English, French and German. This feature highlights the contribution from many European countries involved in preparation and reviews. The ISSMGE recommendations include a detailed explanation of metrological terms such as 'error of measurement'. The main portion of the publication covers detailed procedures for the common geotechnical soil-classification tests, such as particle size distribution and Atterberg limits. The document also describes the consolidated undrained triaxial test and the direct shear test. Supporting activities included detailed analysis of various European methods for determination of liquid limit. The recommendations support two types of cone methods for liquid limit determination. The use of the Casagrande apparatus requires due caution. The input of the ETC5 document in European standardisation is a possible, but long-term process. The discussion session led to a contribution from Germany about funding for further research on laboratory testing, in-situ testing and sampling programmes.

Extra workshop – Sheet piling field test in Rotterdam

Chairman: Prof. F. van Tol (Netherlands)
Secretary: A. Jonker (Netherlands)

The chairman, Prof. F. van Tol, opened this special meeting on the Rotterdam sheet pile field test. In 1993 a full-scale sheet pile wall field test was held in Karlsruhe. In this test a sheet pile wall was brought to an ultimate limit state by simulation of strut failure. The subsoil conditions in this test consisted of sand; the groundwater level was below the bottom of the 5 metre deep excavation.

Since the subsoil conditions in the major part of the Netherlands consists of soft clay and peat with a high groundwater level and because of recent development of Eurocodes for the design practice for steel sheet piling, especially in plastic design and oblique bending, an additional field

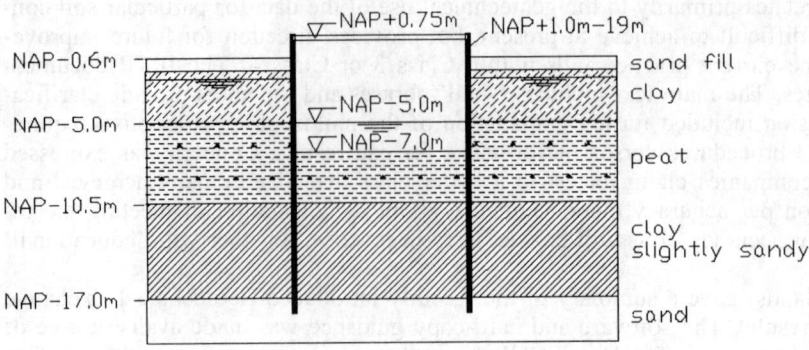

test was required. These developments have led to the initiative for a full-scale field test on two steel sheet pile walls in soft soil at Rotterdam. The main goal of field test was to examine the performance of two facing single strutted steel sheet pile walls with a length of 19 metres and a width of 12 metre. In one wall a plastic hinge was to be developed in order to study a redistribution of bending moments; the other wall was composed of double U-piles to examine the phenomenon of 'oblique bending'. The former wall had to be brought into an ultimate limit state and the latter wall into a serviceability limit state. For the Rotterdam field test a question for type A predictions was distributed to 50 geotechnical engineers world-wide.

For the field test a square cofferdam of approximately 12 by 12 metre was constructed in which the two test walls were included. In order to obtain a benchmark for calculation tools, special measurements were foreseen to obtain a two-dimensional performance of both test walls. A cross-section of the test set-up is presented in the following figure.

The procedure of the field test is subdivided in 4 main stages:
- Dry excavation to NAP-4.0 metre depth;
- Wet excavation to NAP-7.0 metre depth, water level at NAP-1.5 metre;
- Wet excavation to NAP-7.0 metre depth, water level at NAP-5.0 metre;
- Continuation of stage 3 for 6 months.

For the Rotterdam sheet pile wall field test 2 types of predictions are asked for. Predictions of the wall displacements, bending moments, strut forces and earth and water pressure were asked for the above mentioned stages. In the first prediction (prediction type 1), in which a suggestion for soil parameters was provided by the initiators, the aim was to make geotechnical engineers aware of the problems related to plastic design and of friction in the interlocks of U-profiles, since these two phenomena have a large influence on the design and construction of safer and more economic sheet pile walls. For the other prediction (type 2) the whole set of soil investigation was supplied to the predictor. Twenty-one predictions have been submitted, 8 type 1 and 13 type 2. An overview of the predicted bending moment is given in the following table.

Bending Moment (kNm/m)	Number of predictions
< 200	2
200 < M < 250	3
250 < M < 300	1
300 < M < 350	4
350 < M < 400	5
M > 400	7
Plasticity	2

Increase of Bending Moment after 6 months	Number of predictions
0% to 5%	10%
5% to 10%	40%
> 10%	20%
Plasticity before 6th month	30%

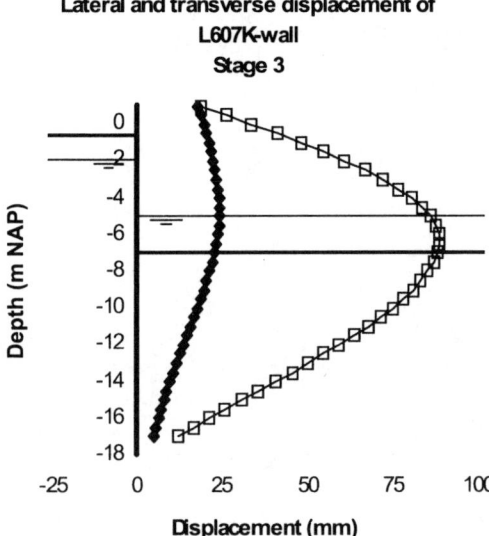
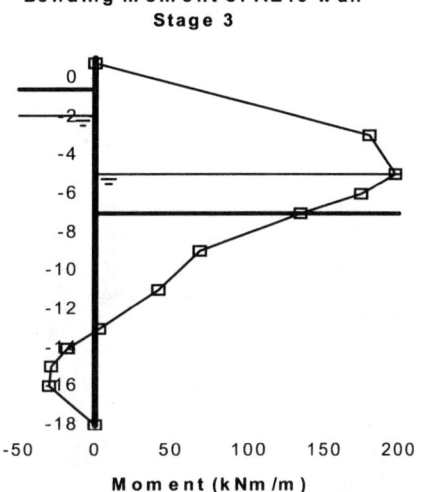

—◆— Transverse displacement —☐— Lateral displacement

A. Kort (Netherlands) presented the first results of the test. The results of stage 3, which had just been achieved – during the conference the test was going on -, showed a maximum bending moment of about 200 kNm/m. This is considerably lower than most of the predicted values and a plastic hinge was not formed at that stage. The next figure shows some preliminary results of the behaviour of both walls. Other typical observations were the decrease of water pressure behind the wall as the wall displaced towards the pit and some evidence of oblique bending.

In the discussion B. Obladen (Netherlands) suggested that arching in the soil is a possible explanation for the observed behaviour of the test walls. Did the initiators consider this aspect?

Prof. van Tol answers that he expects that in this very soft soil, the contribution of arching is not of great importance, because the load on the wall will be dominated by the pore-pressures. W. van Niekerk (Netherlands) asked whether pore pressure cells have been installed inside the test pit. He refers to a job site in Groningen where negative pore pressures in a clay layer were measured after an excavation. For technical reasons it was not possible to install pressure cells inside the pit. However, pore pressure cells have been located on and behind the sheet pile wall.

Although the field test did not behave as predicted, Prof. van Tol considered that the results of the test are interesting and that they will certainly contribute to a better understanding of the behaviour of steel sheet piling in soft soil. The test will continue until the end of 1999. Publication of the final results is foreseen.

Information technology in the geotechnical profession

Full papers of workshop 4

Validation of computer programs in geotechnical design

Andrew Bond
Geocentrix Ltd, Banstead, Surrey, UK

Keywords: software, validation, modelling

ABSTRACT: The Paper discusses what can be done to counter perceived problems of using computer technology for design. It describes the four key components of today's electronic offices: people, software, hardware, and process. The distinction between engineering, conceptual, and computational models is discussed.

1 INTRODUCTION

"Information technology is providing greatly increased power to generate and transfer information ... [but] safeguards are needed to ensure competent use of [this] generated material" (11th Report of SCOSS 1997)

The main worries are (10th Report of SCOSS 1995):
– Users with inadequate engineering knowledge,
– Communication gaps between the engineer and the software developer,
– Programs being used out of context,
– Checking processes that are not sufficiently fundamental,

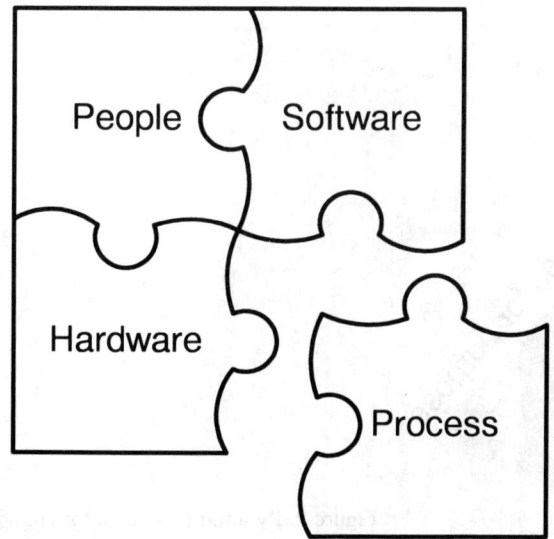

Figure 1. Key components of today's electronic office.

- A program's limitations may not be apparent,
- Programs dealing with unusual structures are particularly weak.

This paper looks at the four key components of today's electronic office (see Figure 1), in an attempt to identify what can be done to alleviate these concerns.

2 SOFTWARE

ISO 9126:1991 defines six characteristics that describe software quality. These are (see Figure 2):
- Functionality – can the software do what is required?
- Reliability – is it well tried and tested?
- Usability – can it be used effectively without significant error?
- Efficiency – does it need excessive computer resources?
- Maintainability – can it be updated if necessary?
- Portability – can it be run on several computers?

The relative importance of the six quality characteristics is indicated by size of each 'stone' in the pyramid of Figure 2.

ISO 9126 recommends evaluating software on a scale of excellent to poor in all of the categories described above (where relevant). Since the fitness of a computer program for a particular purpose depends on that purpose, no general rating levels can be given but should be defined for each specific evaluation.

2.1 Hardware

Software cannot run effectively without the proper hardware:
- Suitable input devices must be available (e.g. keyboard, mouse, digitizer, scanner, microphone, etc.);
- The means of transferring data from other electronic devices must be available (e.g. floppy-disk, CD-ROM, tape, Internet connection, etc.);
- Network connections (if used) must be adequate (e.g. rate of data transfer, capacity of storage media, etc.);
- Suitable output devices must be available (e.g. printer, floppy-disk, CD-ROM, tape, Internet connection, etc.);
- Adequate backup facilities must exist (floppy-disk, CD-W, tape, ZIP-drive, etc).

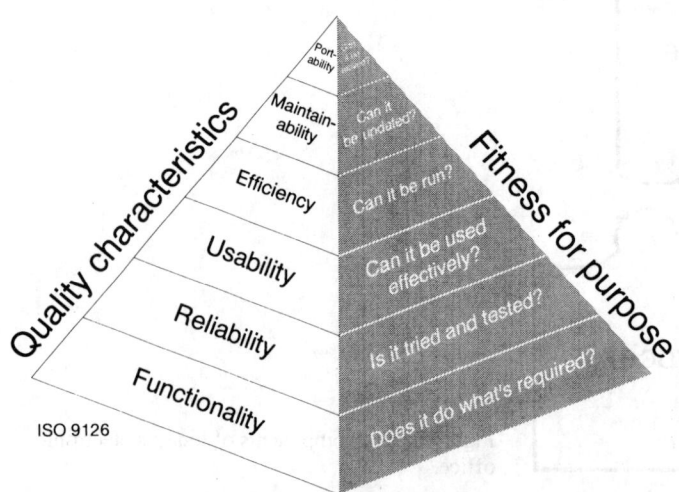

Figure 2. Pyramid of virtues for engineering software.

3 PEOPLE

People are at the heart of the electronic office. The roles played by members of a computer-based project team are illustrated in Figure 3. In a large office, each role in the computer-based project team would be assigned to a different person: in a small office, some roles are combined.

The technical manager of a computer-based project team is responsible for planning the work, obtaining the necessary equipment (hardware and software), and reviewing the outcome. He should be a chartered engineer with sufficient expertise to assess and take responsibility for the overall fitness-for-purpose of the team's work.

The analyst's role within this team is to develop models, run analyses, and to check the results of his work. The qualifications and experience needed by an analyst depend to a great extent on the complexity and usability of the software. Proper training of analysts is essential to ensure their competence, but training is not in itself sufficient – experience also counts for a lot.

The person responsible for checking the results of the modelling exercise should not be the analyst. Successful modelling relies on an independent check of the work being performed either in-house (for example, by a senior engineer) or externally (for example, by a checking agency). On many projects the technical manager also serves as the checker.

Finally, a software specialist is often needed to assist in the selection of suitable software and to run the analyses. The software specialist may be someone from within the organization performing the analysis or an external consultant (possibly an employee of the software vendor).

Each person assigned to a task within the computer-based project team should have sufficient skill and experience to check his work for fitness-for-purpose. In this context, the work is fit-for-purpose if it accurately, adequately, and reliably meets the acceptance criteria previously established for it (see the later discussion of acceptance criteria).

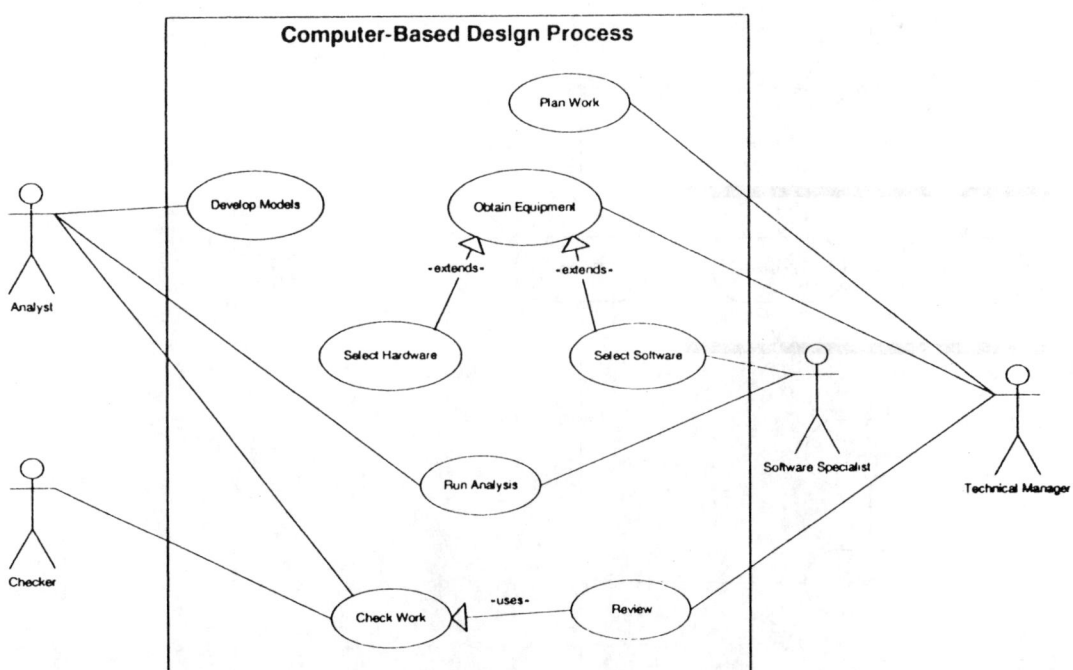

Figure 3. Roles played by members of a computer-based project team (UML use-case diagram).

4 PROCESS

Modelling an engineering structure involves five separate activities, as shown on Figure 4.

In the planning phase, the engineer decides what the purpose of the modelling exercise is, for example:
- To demonstrate that a footing has an adequate factor of safety against excessive settlement;
- To predict the short- and long-term deformation of an embedded retaining wall;
- To estimate the rate of flow of water through an earth dam.

Once the purpose of the exercise has been determined, the engineer produces models of the engineering system, capturing those features of the structure and its surrounds that are relevant to the subsequent analysis.

As a parallel activity, the equipment (hardware and software) needed to study the model is obtained, providing the means for the engineer to perform whatever calculations are necessary to answer the questions posed during the planning phase.

Finally, the engineer reviews the outcome of the modelling exercise and decides whether to accept them as fit for purpose or to reject them - in which case he may revise some aspect of the modelling exercise before re-running the calculations.

The remainder of this paper describes in detail the various activities in the modelling process shown in Figure 4.

The process of modelling an engineering structure involves the construction of a variety of models, including:
- An *engineering model*
- One or more *conceptual models*
- One or more *computational models*

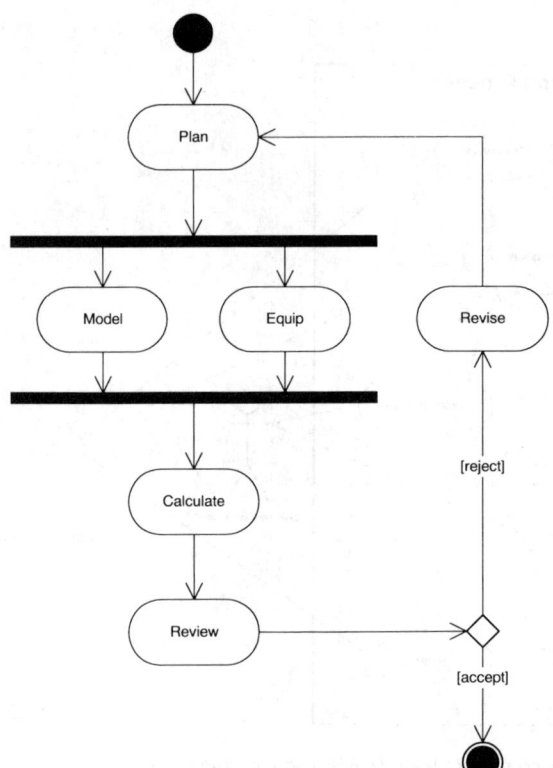

Figure 4. Overview of the modelling process (UML activity diagram).

The *engineering model* is a definition of the physical entity (as exists or as planned) that will be studied in the modelling exercise. The engineering model should be designed to meet the purpose of the modelling exercise. Several different engineering models may be needed to capture the full behaviour of an engineering structure.

A *conceptual model* is a simplified form of the engineering model, which omits physical features that have no significant effect on the aspect of the system being studied. In order to obtain an analytical solution, it may also omit features which are significant.

A *computational model* is a model that allows a solution to the conceptual model to be obtained. Different models are needed for different computer programs.

4.1 *Example of different models*

Figure 5 illustrates the difference between the engineering model of a slope and possible conceptual and computational models for it. The conceptual models (left, circular slip surface; and right, two-part wedge mechanism) capture the physical characteristics of the engineering system and the method of analysing it.

Two possible computational models are shown in Figure 5 for the circular slip mechanisms: one using a smaller slice width than the other. Typically, an analyst would perform a parametric study to determine whether the coarser model (coarser model sacrifices resolution for computational speed) is fit for its intended purpose.

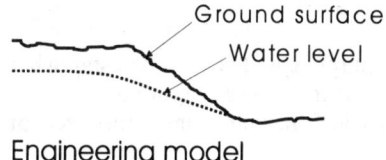

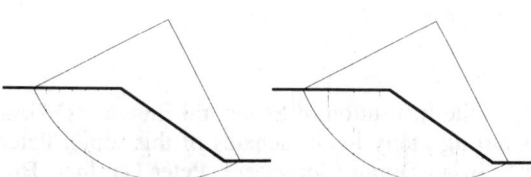

Figure 5. Engineering, conceptual, and computational models for a slope.

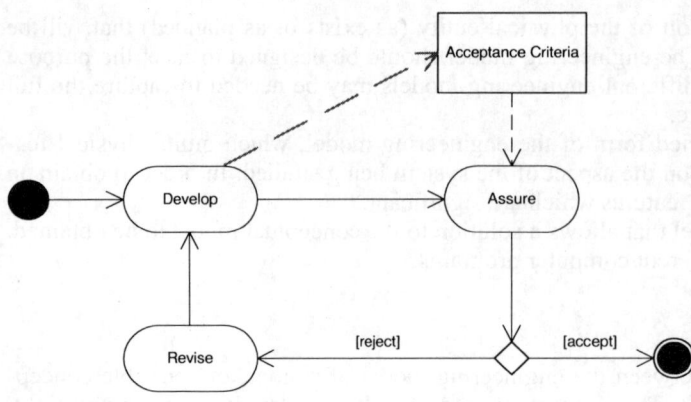

Figure 6. Acceptance criteria act as the link between model development and model assurance (UML activity diagram).

In practice, an analyst would strive to limit the number of models needed to study a particular conceptual model. An example of when more than one computational model is used is to study the influence of mesh size and shape in a finite element analysis.

4.2 *Acceptance criteria*

Developing models involves two complementary activities:
- Development itself (i.e. creation of the model);
- Assurance (i.e. checking the model).

Acceptance criteria are used to check that the model is fit for its purpose and should be established as a by-product of the development activity, before proceeding to assurance (see Figure 6).

By making the establishment of acceptance criteria a distinct action, attention is focussed on creating effective tests of the proposed model.

5 CONCLUSIONS

Successful modelling of engineering structures depends on several factors, not least of which is a methodical and systematic process which identifies the different types of model commonly used as abstractions of real world entities. By making what may appear at first sight to be fine distinctions between these models, the engineer is able to formulate more effective tests of each model and hence reduce the likelihood of errors creeping into the analysis. Acceptance criteria should be produced during model development so that the subsequent quality assurance activities are as objective as can be.

Software should be evaluated on the basis of established characteristics, whose importance will vary from project to project.

People are essential to the successful use of computers in engineering design. They have many roles to play and the skills they need to perform these roles vary greatly.

A longer version of this paper will appear in due course (Bond & MacLeod 1999).

6 ACKNOWLEDGMENTS

The Author would like to thank the other members of the Institution of Structural Engineers' *Task Group on Safer Computing* for their contributions during many lively debates of this topic: Peter Harris (Chairman), Iain MacLeod (Vice-Chairman), Susan Doran (Secretary), Peter Gardner, Bill Harvey, Nigel Knowles, and Brian Neale.

REFERENCES

Standing committee on structural safety 1997. Structural safety 1994-6: Review and recommendations. 11th Report of SCOSS, SETO Ltd, London, UK.
Standing committee on structural safety 1995. Structural safety 1992-4: Review and recommendations. 10th Report of SCOSS, SETO Ltd, London, UK.
ISO 9126:1991. Information technology – Software product evaluation – Quality characteristics and guidelines for their use 1991. International Standards Organization, Bern, Switzerland.
Bond, A.J. & MacLeod, I.A. 1999. A strategy for computer modelling in geotechnical design. Submitted to Geotechnical Engineering, Thomas Telford Ltd, London, UK.

The emergence of information technology: A state of practice report

K.R. MASSARSCH
MCIT AB, Stockholm, Sweden

Keywords: Communication, communication systems, computers, database systems, education, engineering, future, information technology, Internet, networks, software, World Wide Web

ABSTRACT: The paper summarises the state of practice of information technology. Some aspects of its revolutionising influence on society in general, and on engineering in particular, as well as its unprecedented growth are presented. The main components, on which information technology rests, and which have contributed to its rapid development, are presented from an historic perspective. The importance of efficient and unregulated communication systems and that of local and global networks, such as the Internet and the World Wide Web, for electronic communication is illustrated. The roles of computer hardware and software, as well as their applications to the solution of engineering and other problems are presented. Finally, examples are given of how merged networks will provide interactive and almost unrestricted exchange of information. These exciting developments are presently taking place and will affect not only the scientific world, but even more so to society in general.

1 INTRODUCTION

To present a report on the 'State of Practice of the Information Technology (IT)' is an almost impossible task as developments take place at an accelerating pace. New, amazing developments in the electronic sector and their ingenious application take place on a daily basis. What was considered state of the art a few months ago is today state of practice or even out of date. IT has become part of almost everyone's daily life, at work and in the home. Information Technology, which only a few years ago was reserved for 'scientific computer applications' has during the past few years expanded to all sectors of society. The present report must be read in this context and be viewed as a 'flashlight picture' of the situation at the time of preparation of the report in 1999.

Information Technology has, like few other technical developments throughout history, influenced so many aspects of society, on a professional, educational as well as on a private level. While the IT revolution had started already in the early 1980s, few did fully realise its impact.

1.1 *The emergence of Information Technology*

In a general sense, IT includes all forms of information transport by electronic media. However, the term Information Technology can have different meanings, depending on the context in which it is used. In a broad sense, IT includes all communication of electronic information by, and between computers. It includes many different types of applications, of technical, educational, financial/commercial or entertainment nature. However, the most visual and spectacular application of IT takes place on the Internet and in particular on the World Wide Web ('the Web'). The Web is already affecting all segments of society and more than 80% of all activities on the Internet take place on the Web. As recently as in 1995, a mere 14 million people used the Web. This number has since grown to 76 million in 1998 and is estimates to be 130 million by the year 2000. The rate of expansion has been most dramatic during the past 3 years. In 1999, more than 30% of the population of Sweden has had access to the Internet and the Web.

This explosive development is amazing, considering that the Personal Computer (PC) was introduced around 1983, and that as recent as in 1994, IT-companies were not yet listed on the New York Stock Exchange. The largest company listed in 'Fortune 500' was on April 1, 1999 General Motors Corp., with around 600,000 employees and a turnover of 161 billion US$. The stock value was 64 billion US$. Microsoft Inc., the leading IT-company listed on the stock exchange had at the same time only 22,000 employees, a turnover of 14.5 billion US$, but a stock value of 419 billion US$. If stock value reflects future revenue earning capability, then the most significant developments are expected to take place in the IT sector. For example, in Sweden in 1999, IT companies were already the second largest industrial sector, just behind manufacturing and Swedish construction companies were for the first time investing more in IT than in construction machines!

If one applies the experience from past industrial developments, such as the invention of the steam engine, the train, the car or the aeroplane, then the most important developments are still to come. Revenues will be generated by the application and use of IT tools (communication), rather than from the sale of electronic hardware. It is interesting to note that investment in IT-tools usually does not yield direct financial return by reducing costs. The main benefit stems from efficient use of IT-applications, which increase quality, improve exchange of information and thereby increase competitiveness.

1.2 Reasons for the IT revolution

What are the reasons for the unprecedented pace of development in the IT-sector? In the past, values were created by 'limited availability', which created 'demand'. In the IT-world, however, value is created by 'abundance'. Development costs are generally 'one time investments', after which revenues are received from their application without major investment efforts. This is one of the reason why the market share of successful IT systems can grow so rapidly.

Another important aspect is the 'unrestricted expansion' as governments have difficulties to regulate and tax revenues on a national level. The cost for contact is low. Communication is turned from the sender to the receiver. The electronic contact surface and information surfaces are growing rapidly. The following list includes some of the most important reasons for this explosive development:
- The infrastructure of the telecommunication sector (telephone, TV, satellites etc.) has been opened to electronic communication, allowing unrestrained electronic transfer of information across national boundaries,
- Digital storage of all types of information (sound, print, images, video) has become possible,
- Information can be accessed by large groups, the geographic location is of no relevance,
- IT is a searchable medium ('information pull', compared to traditional 'information push', such as TV, newspapers etc.) where the user can select the information he needs or wishes to receive,
- IT is an 'enabling technology' with almost unlimited applications (needs are 'created'),
- Expansion is practically unrestrained, due to low cost and 'unlimited communication' and
- New concepts and applications can be developed without significant investments.

Most experts agree that IT tools and their applications do not increase efficiency or decrease costs on a larger scale. In that meaning, IT investments 'do rarely pay off'. However, the most important benefits derive from IT-applications, with improved quality and increased competence and knowledge by those who use IT efficiently. As IT is an 'enabling technology', it can create demand and needs with the use of communication and information tools. New products and applications can be created with relatively low investments.

1.3 Applications of IT to engineering problems

The main reason for the success of IT is the merger of different technologies, which includes the following components: computer hardware and software, communication and digitisation of information. These will be discussed in the following sections. The IT process, when applied to engineering problems, can be divided into the following steps:

- Data acquisition/digitisation
- Data transmission
- Data storage
- Data processing/analysis
- Communication of obtained information and
- Printed presentation or display of information.

At least some of the following tools are required for the efficient implementation of IT projects:
- Electronic data acquisition systems including sensors or transducers, amplifiers, cables etc. (hardware),
- Digital input of information (measurement data, typed information, images, sound, video etc.)
- Electronic components/computer hardware including terminals, input/output device and data storage and data output,
- Computer software (conversion of data into information, using analysis software, display and presentation),
- Data transmission hardware (cable, telephone or satellite) or
- Data communication software (Internet, Intranet).

The power of IT is that the below listed applications are not mutually exclusive; in fact, the full power of IT is the merger of different technologies and applications. Some examples of IT application are given below:
- Information retrieval from databases (e.g. library systems),
- Visualisation and display of digital information (3-D, dynamic display, video etc.)
- Interactive analysis (exchange and flow of information between different users)
- Electronic journals and publications
- Education and training
- Conferences using IT tools (phone, video, image, e.g. via Internet)
- Computer-aided design, manufacturing, production etc.
- Geographic positioning and guidance systems (GIS)
- Remote measurements, production/quality control (e.g. in real time)
- Supervision of project execution (remote control, 'active design')
- Monitoring of structures or project performance (warning systems)
- Commercial applications (electronic business).

2 COMMUNICATION SYSTEMS

Throughout history, political and social systems which were able to communicate efficiently, have dominated and controlled important developments. The sphere of interest and concern of an individual is determined by the distance to (and from) which information can travel within a few days. Initially, the horse was the primary means of transportation. Societies, which had access to efficient infrastructure and thus information systems, could gain political and social dominance. An excellent example is the emergence of the Egyptian culture more than 5 thousand years ago. The Nile, which both made possible and maintained life in Egypt, was also the principal highway of the land. Until the early New Kingdom (1600 BC) Egyptians lacked wheeled conveyances and the horse. The Nile offered the obvious means for travel and transport, which was considerable from early on. Travel on the Nile was facilitated by the direction of the prevailing wind, which with almost unceasing regularity blows from the north. A boat travelling south, against the flow of the stream, could use sails, one travelling north could proceed with the current. The river was used to such an extent that the regular words for 'go north' and 'go south' were determined by hieroglyphic script by boat. Go north had a boat with no sails and go south had a boat with sails.

It is apparent that communication is the most important prerequisite for sustainable political, social and technological development. The technology of modern communication results from the confluence of many types of inventions and discoveries, some of which (the printing press, for in-

Figure 1. Sailboats on the Nile have provided and still provide efficient transportation and communication.

stance) actually preceded the industrial revolution. Inventions of the 19th and 20th centuries have made available newer means of communication, particularly broadcasting, without which the present near-global diffusion of printed words, pictures, and sounds would have been impossible.

2.1 Development of data communication systems

Electronic information systems are a phenomenon of the second half of the 20th century. Their evolution is closely tied with advances in two basic technologies: integrated circuits and digital communications. Advances in the design of these chips, which were first developed in 1958, are responsible for an exponential increase in the cost/performance ratio of computer components. Full exploitation of these developments for the information systems requires comparable advances in software. Their major contribution has been to open the use of computer technology to persons other than computer professionals. Interactive applications in the office and home have been made possible by the development of easy-to-use software products for the creation, maintenance, manipulation, and querying of files and records. In 1968 the Federal Communications Commission (FCC) allowed the direct connection to the telephone network of non-telephone devices. This milestone decision created a new data communication applications. The main components of data communication are: *networks, terminals, protocols and software*. These will be discussed in the following section.

2.1.1 Electronic Networks

The full potential of data communications rests in the application of networks, where a remote terminal is connected to a computer point-to-point. When there are several terminals at a remote location, requiring connection to the computer, it is possible and less costly to use only the one line in combination with a 'multiplexing' (line-sharing) device. As terminals from different remote sites require access to the centrally located computer, the network grows in complexity. Waystations, called nodal processors or 'nodes', become necessary within the network to interact with the terminals and the computer, storing, forwarding, and controlling the networks data flow. As more computers and terminals are added, the network design becomes even more complex to provide the capability for any terminal to reach any computer or any other terminal.

The user can work on a private network, which yields many control and security benefits, or a public network to move data. Among the different types of network designs is circuit-switching, which involves a straightforward transfer of data from one path onto another. Message-switching, on the other hand, adds a store-and-forward function: the message is stored at a network node, possibly checked for errors, then transferred onto another circuit (path). A popular variation of the message-switched network is the packet-switched network. Here, as the message enters the network, it is divided into discrete segments, called packets. Each packet travels the network independent of the others, with suitable identification. A packet is routed from node to node, until it reaches its final destination. There, all the packets are reassembled in their proper order, and presented to the destination device as the entered message.

The local area network (LAN) is commonly used for data communication and consists of a loop of coaxial cable, interconnecting offices in adjacent buildings, operating at megabit/second data rates. Ideally, the LAN enables the user to attach any communicating office device to the loop, automatically gaining access to all the other devices so connected. An example is the Ethernet where each office terminal device may readily access common databases, rather than having to maintain its own. With the advent of digitised voice transmission, the office network has become the common shared medium for all office transmissions, thus avoiding the need for separate, costlier facilities. The increasing application of optical fibre cable and its 100 Mbit/s (and up) data rates has led to newer LANs and developing standards. One offshoot, the metropolitan area network (MAN), has led to the fibre distributed data interface (FDDI).

To enable resource sharing on a global scale, by allowing LANs to connect to each other, an 'Internetwork interface', called a gateway, is needed. It has the responsibility for end-to-end accountability, data routing, and traffic flow control. With gateway availability, LANs interconnect and also gain access to existing long-haul networks. In wide-area networks, such as the Internet, which connect thousands of computers around the globe, computer-to-computer communication uses a variety of media as transmission lines, such as electric-wire audio circuits, coaxial cables, radio and microwaves (as in satellite communication), and, most recently, optical fibres. The latter are replacing coaxial cable in the Integrated Services Digital Network (ISDN), which is capable of carrying digital information in the form of voice, text, and video simultaneously.

It can be shown that every 2-4 years, Ethernet bandwidth is increasing by an order of magnitude. Applied to conventional communication problems, this implies that the number of highway lanes between major cities would increase from 2 to 2000 lanes! Also, every 12 months, the density of light wavelengths in a single strand of fiber is doubling.

One data communications technology that is fast maturing is lightwave transmission. The medium in greatest use is glass, in the form of optical fibres. Using a light-emitting diode (LED) or a laser as the light source, data rates in the multimegabit-per-second range have been achieved. Other advantages are the material's ready availability, light weight (lighter than copper), and narrow gauge (less than coaxial cable), all made available in a sufficiently strong packaged product. Connectors and field-splicing techniques are available.

An important element of the data communications network is the satellite – particularly, the geostationary satellite. It acts as a relay and disseminating station for all forms of electrical signals. When a network requires satellite use, 'dish' antennas may be positioned on building roofs, to minimise land-line use. A drawback in the application of satellites to data transmission is the inherent propagation delay of this technique. There are several schemes to control the delay's effect, especially noticeable in interactive (inquiry-response) applications. The user must balance their higher costs against those of transmission via cable and microwave links. For non-interactive transmission, such as batch data, video, and facsimile, the delay is often acceptable.

2.1.2 Terminals

The most universal data communications device is the terminal, which commonly has a typewriter-like keyboard, and a cathode-ray-tube (CRT) screen or a hard-copy capability. The terminal permits the user to gain access to a wide range of databases (information sources), including business, training, entertainment, and interactive applications. The terminal can be a multifunctional device: a microcomputer or personal computer (PC). Depending on the amount of intelligence, it can not only handle interoffice message communications, such as electronic mail, but also perform data processing. With today's integrated-circuit chip technology, the PCs include microprocessors, each with its own designated function. Only cost and the abilities of its programmers limit the information processing power. One 'older' device is the minicomputer. As a communications processor, it 'front-ends' a mainframe computer to handle the network traffic, thus freeing its host for the 'number-crunching' tasks it handles more efficiently. The communications processor also functions as a switch, connecting terminals to each other as required. When a user desires access to the host's database, the front end opens a path to the mainframe.

Facsimile terminals are also evolving. New digital techniques enable scanning in seconds the material to be transmitted. Among these techniques is the terminal scanner's ability to skip over or 'compress' redundant characters, including 'white space'. Of course, the receiving machine interprets the received compression codes and restores the original material in its entirety. Adding a faxboard to a PC adds facsimile-sending capability.

2.1.3 *Protocols and Software*

Data communications devices communicate with each other using protocols. They range from rules for simplex (one-way-only) transmission to packet-network, high-level data link control. The more complex the protocol, the more intelligence required of the terminal/computer device and of the network.

The increasing networking application of PCs has given rise to PC-specific programs for controlling and communicating over a variety of networks. In many instances, PCs have replaced terminals and other networking computers. A software-related function of increasing importance to the network planner and user is database management. Database management systems provide users with a method of readily accessing data no matter where the data resides in a network. These services are usually available on a timeshared basis, meaning accessible to many users 'simultaneously' (i.e., it appears so to the user). The aim is to approach the ideal concept of totally distributed network processing. The user accesses a database by subject, and the network connects the user terminal to the proper computer. The user is completely unaware of this network operation.

Enhanced electronic document interchange (EDI) services provide computer-to-computer links and high-speed document communications. The network interface unit is often microprocessor-equipped to manage data-packet processing, circuit connection, and error detection. Another notable development is the digital private branch exchange, which can handle both data and voice by digitising the voice signals. Digitised information from PCs, facsimile equipment, and other devices are all tunnelled through the digital exchange. Its own memory and switching functions, and interfaces to analogue and digital networks (including packet-switched types), are all made possible by the application of microprocessor and chip technology.

Speech recognition and voice response technologies are being applied to data communications devices in several areas. One is voice mail, where a computer synthesises messages for the user who dials up a 'mailbox'. Another is the voice-activated workstation (both with and without a data communications interface). Evolving techniques of natural-language processing and understanding, knowledge representation, and neural process modelling have begun to join the more traditional repertoire of methods of content analysis and manipulation. The use of these techniques opens the possibility of eliciting new knowledge from existing data, such as the discovery of a previously unknown medical syndrome or of a causal relationship in a disease.

3 THE EVOLUTION OF THE INTERNET

The Internet had its origin in a U.S. Department of Defense programme, called Arpanet (Advanced Research Projects Agency Network), established in 1969. The aim was to provide a secure and survivable communications network for organisations engaged in defence-related research. When one machine was cut off, as long as the message could get out to one machine, it would go around and make its way somehow to the receiver.

The National Science Foundation (NSF), which had created a similar and parallel network called NSFNet, took over much of the communication technology from Arpanet and established a distributed network of networks, capable of handling far greater traffic. Researchers and academics in other fields began to make use of the network, the Usenet. DEC Corp. was the first company to subsidise communication on the Usenet, to enhance communication of their employees. While users of the Arpanet had to have government permission, the Usenet was designed to let all people connect. Networks work best when one can connect to the edges. There is no centre or point of control. As the network architecture had no central control, the information was not censurable.

Gradually, Usenet and Arapanet turned into Internet. The Internet is a global network, which connects many computer networks. While networks connect individual computers, which have information on each machine, the Internet connects information independent of the computer or the operating system. It is based on a common addressing system and communications protocol called TCP/IP (Transmission Control Protocol/Internet Protocol). TCP/IP is a protocol standard of data packet transmission which has been incorporated in the UNIX operating system.

As the Internet grew, networking was changing from a research tool to a social phenomenon. Suddenly a broader range of people was drawn in from 1983-4. The original uses of the Internet were electronic mail, file transfer (ftp, or file transfer protocol), bulletin boards and newsgroups, and remote computer access (Telnet). With the emergence of modems, computers could connect to other computers and communicate and talk on Bulletin Boards. These were on the fringe of the Internet. All that was required was a phone line and a computer connected to a modem. Individuals could 'meet' and share information on specific topics of interests. Many commercial computer network and data services also provided at least indirect connection to the Internet.

An important development was the introduction of indexing tools on the Internet, such as gopher in 1990. A tremendous increase in Internet traffic resulted from this innovation. By the mid-1990s the Internet connected millions of computers throughout the world. The number of users of the Internet is growing by about 80 millions per year.

4 THE WORLD WIDE WEB

The World Wide Web is a simple yet gigantic innovation,. It consists of three parts, which extend beyond the ftp and gopher: a new concept, the hypermedia browser; a new standard for hypertext files, HTML; and a loose convention for multimedia file types (image, sound, video). The World Wide Web (WWW) enables simple and intuitive navigation of Internet sites through a graphical interface. It expanded dramatically during the 1990s to and has become the most important component of the Internet. Today the emergence of the WWW is driving the explosion of the Internet. The WWW is a self-organising structure within the Internet, and consists of the following three elements:
– Hypermedia browser,
– New standard for hypertext files HTML and
– Loose conventions for multimedia file types (image, sound, video).

4.1 Browser

A browser displays hypertext files. Berners-Lee at CERN developed the first browser in 1989. He wrote a programme without a hierarchical structure, developing a web of interconnected relationships. This text-based Web browser was made available for general release in January 1992. The hypertext program introduced a completely new concept: to link an idea on one computer to an idea on another computer. Suddenly, it became possible to jump between ideas from one computer to another computer. Thereby, group knowledge would become accessible to an unlimited number of people at remote locations.

The first graphical browser (called Mosaic) was developed by Andreeasen at the National Center for Supercomputing Applications (NCSA) during 1992 and released in 1993. Besides displaying hypertext is also launched other programmes to display images, sound and movie files.

4.2 Hypertext

Hypertext is a text file with hotlinks, which initiates a jump of the mouse cursor to a different location in the file, or to a different file. A hypertext document with its corresponding text and hyperlinks is written in Hyper Text Markup Language (HTML) and is assigned an online address, called a Uniform Resource Locator (URL). HTML format is a standard way of indicating hotlinks and their destinations, as anchors. Also images may be 'hot'; that is clicking on them results in the jump to another file. HTML is the second piece that is needed to build the WWW.

4.3 Conventions for multimedia

Hypertext links between different parts of a document or between different documents create a branching or network structure that can accommodate direct, unmediated jumps to pieces of related information. The tree-like structure of hyperlinked information contrasts with the linear structure of a printed encyclopaedia or dictionary, for example, whose contents can be physically accessed only by means of a static, linear sequence of entries in alphabetical order. Hypertext has been used most successfully by the interactive multimedia computer systems that came into commercial use in the early 1990s.

The WWW is the leading information retrieval service of the Internet and gives users access to a vast array of documents that are connected to each other by means of hypertext or hypermedia links. The Web operates within the Internets basic client-server format. Servers are computer programs that store and transmit documents to other computers on the network when asked to, while clients are programs that request documents from a server as the user asks for them. Browser software allows users to view the retrieved documents. The incorporation of pictures, moving images as well as sound is changing the Web again.

4.4 Electronic commerce

The interactive aspect of the WWW offers powerful communication links between different groups, which wish to sell or purchase goods or information. This has led to the rapidly growing e-commerce. Financial transfers (digital business) will become even more important in the future with the introduction of 'electronic cash'. Wallet money can be stored in digital form on the computer can become part of the electronic data exchange/transfer.Kolla denna mening!

4.5 Viruses and secrecy

Hackers exploit the loopholes of the network, by getting into organisations or peoples computers and systems. Viruses started a new problem, that of access and secrecy. Access by dialing up a computer. In 1988 the first 'Internet worm' was released which entered computers via e-mail. It turned out that the Internet was completely unprepared and wide open to a college graduate, who entered government computer networks. This was the single most significant incident, which gave the industry a wake up call.

5 EVOLUTION OF COMPUTERS

The earliest known analogue computing devices were sticks or ropes used specially for measuring length. The Abacus frame was probably the first calculation machine, which had its origin in China around 2000-500 BC. In 1400 the first slide rule was developed which could be used for all four calculation methods. Mechanical calculating machines were invented in Europe during the 17th century. In Germany, Schickard, a friend of the astronomer Kepler, invented around 1623 the first mechanical calculator. The records were lost, however, during the Thirty Years' War. Pascal (1623-1662) constructed the first calculation machine, developed for tax collection, used cogwheels to drive the machine. This machine could only perform addition and subtraction. 30 years later, the German mathematician and philosopher von Leibnitz (1646-1716) developed a machine, which also could perform multiplications and divisions. Another important invention which influenced computer technology was by Newton (1643-1727) who among others developed differential calculus and integration which are foundations of modern mathematics.

The industrial revolution required more advanced production systems. The first automated weaving machine was invented by Jacquard, which was programmed using punched cards. Thereby, the first programming concept was thus introduced. Babbage (1792-1871) constructed the first Difference Machine, building on Jacquard's invention. The idea of mechanically calculating mathematical tables first came to Babbage in 1812 or 1813. Later he made a small calculator that could perform certain mathematical computations to eight decimals. This invention can be considered the prototype of a digital computer. He later developed the Analytical Engine, which was programmed using punch cards, had memory and could thus be programmed. The first commercially available calculator, the 'arithmometer', was produced in 1820 by de Colmar of France.

5.1 Electronic computers

Hollerith is the inventor of a tabulating machine that was an important precursor of the electronic computer. By the time of the census of 1890, he had invented machines to record statistics by electrically reading and sorting punched cards that had been numerically encoded by perforation position. The invention was a success in the United States but drew much more attention in Europe, where it was widely adopted for a number of statistical purposes. In 1896, Hollerith organised the Tabulating Machine Company, incorporated in New York, to manufacture the machines; through subsequent mergers it grew into the International Business Machines Corporation (IBM).

The transistor circuits inside modern electronic digital computers perform such logic operations. A significant step in the development of analogue computers was the American engineer Bush's invention of a device for amplifying the small torque generated by a rotating wheel. In 1930 Bush and his colleagues at the Massachusetts Institute of Technology constructed a mechanical analogue computer, called the differential analyser, for solving differential equations.

Zuse in Germany was undertaking Work on calculating machines, and during the period 1936-49 he developed a series of four ZUSE computers, which were electromechanical relay machines. In 1939, Aiken began designing a fully automatic large-scale calculator and in 1944 put into operation the Automatic Sequence Controlled Calculator, commonly known as the Harvard Mark I. Data were entered on punched cards, and output was recorded either on punched cards or by an electric typewriter.

It had been thought for more than two decades that the first electronic digital computers were the Colossus built in England in 1943 and the Electronic Numerical Integrator and Computer (ENIAC) built in the United States in 1945. However, the first electronic digital computer was actually built by J.V. Atanasoff, an American theoretical physicist. During the period 1937-42, he built two small-scale, special-purpose electronic computers. Atanasoff's computer greatly influenced the future of electronic computing technology. It was the first machine to use electronic means to manipulate binary numbers. Several concepts introduced by Atanasoff remain important in today's computers, including the use of capacitors in dynamic random-access memories, the regeneration of capacitors, and the separation of memory and processing.

A major breakthrough in the development of electronic computing came when Eckert, and his colleagues, built the high-speed electronic digital computer, known as the ENIAC. Completed by February 1946, the ENIAC was the first general-purpose electronic digital computer. It contained roughly 18,000 vacuum tubes and measured about 2.5 m in height and 24 m in length. The machine was more than 1,000 times faster than its electromechanical predecessors and could execute up to 5000 additions per second.

5.2 Modern computers

Modern computers are sometimes classified into generations, with each generation characterised with some important technology. Computers of the first generation employed vacuum tubes as logic gates. They were the first widely successful commercially available computers. The UNIVAC I was the first commercial computer in the United States; 46 machines were sold.

The second generation began when the first successful commercial computers employing transistors were introduced by the Control Data Corporation and IBM around 1960. Although it was invented in 1947, the transistor did not become commercially available as a viable alternative to the vacuum tube until about 10 years later, after a series of improvements had been made. By using transistors in electronic circuits, along with an improved magnetic-core memory, computer manufacturers were able to produce digital computers that were smaller, faster (capable of executing up to 100,000 instructions per second), less expensive, and more reliable than those available before.

During the late 1960s and the 1970s, transistor circuits, consisting of hundreds or thousands of transistors, diodes, resistors, and the connections among them, were further miniaturised and placed on a rectangular silicon chip of less than 6.5 millimetres (0.25 inch) on each side. Transistor circuits integrated on tiny silicon chips led to dramatic advances in computer technology. The use of such integrated circuits (ICs), which characterised the third generation, permitted the construction of computers with higher speeds and reliability at lower cost.

With continuous progress in integrated-circuit fabrication technology, the integration size of an IC chip was improved through large-scale integration (LSI), which made it possible to pack thousands of transistors and related electronic components onto a single chip.

The integrated circuit had all the arithmetic, logic, and control circuitry necessary to function as a central processing unit (CPU). The first commercially successful microprocessor chip, 8080, was introduced by the Intel Corp. in 1974. The manufacturing of random-access memory (RAM) chips using transistors also became feasible owing to the progress of LSI. The microcomputer (with microprocessor, memory, and many other IC chips, as well as a cathode-ray tube, a keyboard, and an I/O unit) became a forerunner of the desktop computer.

In the early 1990s millions of transistors and related electronic components were squeezed onto each tiny IC chip. The computers of the 1980s employing LSI and VLSI technologies have often been referred to as fourth-generation, although the differences between them and third-generation machines are not very clear.

The fifth-generation computer project emphasised artificial intelligence, focusing on such matters as machine reasoning and logic programming languages. The fifth generation has not yet had any direct significant impact to date on the commercial computer market.

5.3 Personal computers

Computers small and inexpensive enough to be purchased by individuals for use in their homes first became feasible in the 1970s, when large-scale integration made it possible to construct a sufficiently powerful microprocessor on a single semiconductor chip. The personal computer industry began in 1977, when Apple Computer, introduced the Apple II, one of the first pre-assembled, mass-produced personal computers. Radio Shack and Commodore Business Machines also introduced personal computers that year. In 1981, IBM introduced the Personal Computer, or IBM PC. The IBM PC became the world's most popular personal computer, and both its microprocessor, the Intel 8088, and its operating system, which was adapted from the Microsoft Corporation's MS-DOS system, became industry standards.

In 1985 the Microsoft Corporation introduced Microsoft Windows, a graphical user interface that gave MS-DOS-based computers many of the same capabilities of the Macintosh. Windows became the dominant operating environment for personal computers.

By 1990 some personal computers had become small enough to be completely portable; they included laptop computers, notebook computers and pocket, or palm-sized, computers. Multimedia personal computers equipped with CD-ROM players and digital sound systems allow users to handle animated images and sound (in addition to text and still images) that are stored on high-capacity CD-ROMs. Personal computers were increasingly interconnected with each other and with larger computers in networks for the purpose of gathering, sending, and sharing information electronically.

Present-day computers are remarkably versatile machines capable of assisting humans in nearly every problem-solving task that involves symbol manipulations. Television, on the other hand, has penetrated societies throughout the world as a non-interactive display device for combined video and audio signals. The impending convergence of three digital technologies, namely, the computer, very-high-definition television (V-HDTV), and ISDN data communications is all but inevitable. In such a system, a large-screen multimedia display monitor, containing a 64-megabit primary memory and a billion-byte hard disk for data storage and playback, would serve as a computer and, over ISDN fibre links, an interactive television receiver.

5.4 *Microprocessor technology*

This multitude of new products and capabilities has been made possible by the tremendous progress of microprocessor technology. Because of the advances in this area, personal computers have become more powerful, smaller, and less expensive, which has enabled computer networks to proliferate. Many of the tasks that were traditionally performed by mainframes have been transferred to personal computers connected to communications networks. Although the mainframe continues to be produced and serves a useful purpose, it has been used more often as one of many different computers and peripheral devices connected to computer networks.

For more than two decades the capacity of the basic integrated circuit, the dynamic random-access memory (DRAM) chip, has doubled consistently in intervals of less than two years: from 1,000 transistors (1 kilobit) per chip in 1970 to 1,000,000 (1 megabit) in 1987, 16 megabits in 1993, and 1,000,000,000 (1 gigabit) predicted for the year 2000. The speed of microprocessor chips, measured in millions of instructions per second (MIPS), is also increasing near-exponentially: from 10 MIPS in 1985 to 100 MIPS in 1993, with 1000 MIPS predicted for 1995. By the year 2000 a single chip may process 64 billion instructions per second.

If in a particular computing environment in 1993 a chip supported 10 simultaneous users, in the year 2000 such a chip could theoretically support several thousand users.

6 EVOLUTION OF ENGINEERING SOFTWARE

Software is an essential part of information technology application. In the following sections, the evolution of engineering concepts with respect to design and implementation of IT-projects will be discussed.

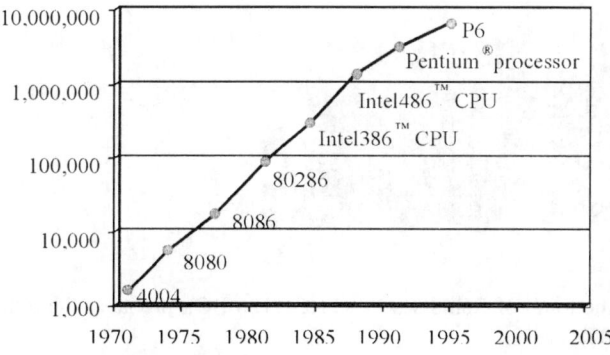

Figure 2. Increase of processor and computing power, note the logarithmic scale (in CPU MIPS).

6.1 History of design concepts

The history of engineering and architecture may be read in the progressive changes in the solution of structural problems. Three important steps can be identified: 1. The first time a building plan was used for construction purposes, 2. The introduction of perspective (third dimension) and 3. The transfer of drawings into digital form, which makes computer analysis and presentation possible.

Already early in the history of man, major engineering structures were constructed, but primarily for military or religious reasons. As most projects were planned and executed on a trial and error basis, the experience of the builders was of major importance. Probably the earliest remaining engineering documents is a plan of the tomb of Pharaoh Ramesses IV painted on stone was actually found abandoned in the tomb. A second plan, which is incomplete, is written on papyrus and depicts part of the tomb of Ramesses IV. Once the site had been chosen and the plan been drawn up, the workman began to cut the tomb out of the solid rock.

Religious buildings form the majority of surviving works of architecture. All of them were symbolic as well as functional. The symbolism is expressed in the siting and design of temples and in the decoration of walls. The theoretical orientation of most temples is east west, based on the Nile. Many of the monumental works turned into or were engineering problems. Moving and carving stone involved special techniques and craftsmanship. This required construction of roads and ships and extensive earthworks for the final siting. The blocks were taken by the most practical rout to the nearest point on the river, from there they were transported by water on barges as close as possible to the building site. Rough architectural plans of buildings on both papyrus and ostracon (shells) have survived.

Greek architecture also formalised many structural and decorative elements into classical orders, which have influenced architectural style since that time. The Romans exploited the arch, vault, and dome and made broader use of the load-bearing masonry wall; and in the late medieval period the pointed arch, ribbing, and pier systems gradually emerged. At the beginning of the Italian Renaissance, early in the 15th century, the mathematical laws of perspective were discovered by the architect Brunelleschi. He worked out some of the basic principles, including the concept of the vanishing point, which had been known to the Greeks and Romans but had been lost. It is interesting to note that the first recorded patent for an industrial invention was the one granted in 1421 in Brunelleschi. The patent gave him a three-year monopoly on the manufacture of a barge with hoisting gear used to transport marble.

During the past centuries great advances have been made in the human capability to record, store, and reproduce information. The invention of printing from movable type in 1450, followed by the development of photography, teleprinting and telephony had a significant influence on information transfer in engineering. However, the development of methods for transfer of analogue into digital information has opened completely new possibilities. This has opened completely new possibilities for the application of IT in engineering design and architecture, as will be discussed later.

Figure 3. A fragmented plan on papyrus of the tomb of Ramesses IV (1163-1156 BC), probably the oldest preserved engineering drawing, (Bierbrier 1982).

6.2 Evolution of computer software

Of great significance in the evolution of the modern digital computer was the work of the English logician Boole. He published in 1847 a theory of logic operators, which became the basis of what is now known as Boolean algebra and the foundation of computer software structure. The first 'computer programme' was written by A. Lovelace for Babbage's automated weaving machine. She can be called the first programmer. In 1945 the Hungarian-born mathematician von Neumann and co-workers proposed that a program be electronically stored in binary number format in a memory device.

Software contains the instructions, which that tell a computer what to do. Software comprises the entire set of programs, procedures, and routines associated with the operation of a computer system. A set of instructions that directs a computer's hardware to perform a task is called a program, or software program. The two main types of software are system software and application software. System software controls a computer's internal functioning, chiefly through an operating system and also controls such peripherals as monitors, printers, and storage devices. Application software, by contrast, directs the computer to execute commands given by the user and may be said to include any program that processes data for a user. Application software thus includes word processors, spreadsheets, database management, inventory and payroll programs, and many other 'applications'. A third software category is that of network software, which co-ordinates communication between the computers linked in a network.

Software is typically stored on an external long-term memory device, such as a hard drive or magnetic diskette. When the program is in use, the computer reads it from the storage device and temporarily places the instructions in random access memory (RAM). The process of storing and then performing the instructions is called 'running', or 'executing', a program. By contrast, software programs and procedures that are permanently stored in a computer's memory using a read-only (ROM) technology are called firmware, or 'hard software'.

Computer programs, the software that is becoming an ever-larger part of the computer system, are growing more and more complicated. The software-engineering process is usually described as consisting of several phases, general consisting of 1. Identification and analysis of user requirements, 2. Development of system specifications (both hardware and software), 3. Software design, 4. Implementation (actual coding), 5. Testing, and 6. Maintenance. Attempts have been made since the early 1980s to build increasingly sophisticated tools to aid the software developer and to automate as much as possible the development process.

Figure 4. 'The Annunciation', fresco showing perspective structure, by Fra Angelico, 1438-45; in the Museum of San Marco, Florence (Encyclopaedia Britannica Multimedia Edition 1999).

6.3 Application of computer software

Computers have been used since the 1950s for the storage and processing of data. A computer provides only temporary storage; any data stored in main memory is lost when the power is turned off. For the permanent storage of data, one must turn to auxiliary storage, primarily the magnetic media such as tapes or disks. Data is stored on such media but must be read into main memory for processing. A major goal of information-system designers has been to develop software to efficiently locate specific data on auxiliary storage and read it efficiently into main memory for processing. The underlying structure of an information system is a set of files stored permanently on some secondary storage device. The software that comprises a file management system supports the logical breakdown of a file into records.

The high-speed digital computer, together with its peripheral equipment, provides an extremely efficient means of analysing, manipulating and modifying stored data. As a consequence, information systems based on such computers are able to carry out diverse tasks, such as: scientific and engineering calculations, translate technical material from one natural language to another, conduct searches of bibliographic literature, provide tutorial instruction in various subjects, assist in design and manufacturing activities, or make decisions for solving complex non-numerical problems. It is interesting to note that the first engineering application of computer software was the solution of the wave equation according to the well-known Smith model in the early 1950s.

The U.S. Department of Defense encountered the problem of supporting hundreds of programming languages and introduced the standard language Ada to reduce software development and maintenance costs. The Defense Department required its new programs to be written in Ada. This programme language was complex and expensive to use, because it had many functions with large memory requirements. Thus, computer users in the civilian market have been choosing standard operating systems instead of programming languages.

A common operating system for different computers is called an open system. In the past, computer manufacturers had maintained their respective proprietary operating systems to be different from others. The open system and also the open architecture described in the following paragraphs are having tremendous impacts on computer manufacturers and software companies in the personal computer market.

IBM adopted the open system strategy when it introduced its PC in 1981 and achieved great success by 1984. Ironically, the explosive growth of PC-compatibles (clones) manufactured by other companies also has led to a decline in IBM's market share of personal computers since 1984.

Application programmes are becoming increasingly complex as new features are continually added. Thus, the development of new application programs and the improvement of old ones are becoming extremely time-consuming. In order to solve this problem, many companies are adopting object-oriented programming. Programmers can increase productivity by incorporating reusable objects instead of writing programs from scratch.

6.4 Database systems

During the early 1960s computers were used to digitise text for the first time; the purpose was to reduce the cost and time required publishing two American abstracting journals. Advances in computer storage, telecommunications, software for computer sharing, and automated techniques of text indexing and searching fuelled the development of an on-line database service industry. Meanwhile, electronic applications to bibliographic control in libraries and archives have led to the development of computerised catalogues and of union catalogues in library networks. They also have resulted in the introduction of comprehensive automation programs.

The explosive growth of communications networks after 1990 has accelerated the establishment of the 'virtual library'. At the leading edge of this development is public-domain information. Residing in thousands of databases distributed world-wide, a growing portion of this vast resource is now accessible almost instantaneously via the Internet. Internet resources of electronic information include selected library catalogues, collected works of the literature, some abstracting journals, full-text electronic journals, encyclopaedias, scientific data from numerous disciplines, software archives, demographic registers, daily news summaries, environmental reports, and prices in commodity markets, as well as hundreds of thousands of electronic-mail and bulletin-board messages.

The vast inventory of recorded information can be useful only if it is systematically organised and if mechanisms exist for locating in it items relevant to human needs. Early file systems were always sequential, meaning that the successive records had to be processed in the order in which they were stored, starting from the beginning and proceeding down to the end. This file structure was appropriate and was in fact the only one possible when files were stored solely on large reels of magnetic tape and skipping around to access random data was not feasible. Sequential files are generally stored in some sorted order (e.g., alphabetic) for printing of reports (e.g., a telephone directory) and for efficient processing of batches of transactions. Banking transactions (deposits and withdrawals), for instance, might be sorted in the same order as the accounts file, so that as each transaction is read the system need only scan ahead (never backward) to find the accounts record to which it applies.

When so-called direct-access storage devices (DASD) were developed (primarily magnetic disks), it became possible to access a random data block on the disk. Files can then be indexed so that an arbitrary record can be located and loaded into the main memory. Since indexes might be long, they are usually structured in some hierarchical fashion and are navigated by using pointers, which are identifiers that contain the address (location in memory) of some item. File systems making use of indexes can be purely indexed, in which case the records need be in no particular order and every individual record must have an index entry that points to the record's location, or 'indexed-sequential'. In this case a sort order of the records is maintained as well as the indexes, and index entries need only give the location of a block of sequentially ordered records. Searching for a particular record in a file is aided by maintaining secondary indexes on arbitrary attributes as well as a primary index on the same attribute on which the file is sorted.

An indexed-sequential file system supports not only file search and manipulation commands of both a sequential and index-based nature but also the automatic creation of indexes.

The database has become a central organising framework for many information systems, taking advantage of the concept of data independence, which allows data sharing among diverse applications. Database management system (DBMS) software today incorporates high-level programming facilities that do not require one to specify in detail how the data should be processed. The programming discipline as a whole, however, progresses in an evolutionary manner. Whereas semiconductor field advances are measured by orders of magnitude, the writing and understanding of large suites of software that characterise complex information systems progress more slowly. The complexity of the data processes that comprise very large information systems has so far eluded major breakthroughs, and the cost-effectiveness of the software development sector improves only gradually.

6.5 Computer visualisation

Computer visualisation, a new field that has grown expansively since the early 1990s, deals with the conversion of masses of data emanating from instruments, databases, or computer simulations into visual displays – the most efficient method of human information reception, analysis, and exchange. Related to computer visualisation is the research area of virtual reality or virtual worlds, which denotes the generation of synthetic environments through the use of three-dimensional displays and interaction devices. A number of research directions in this area are particularly relevant to future information systems: knowledge-based world modelling; the development of physical analogues for abstract quantitative and organisational data; and search and retrieval in large virtual worlds. The cumulative effect of these new research areas is a gradual transformation of the role of information systems from that of data processing to that of cognition aiding.

80 K.R. Massarsch

7 OUTLOOK IN THE FUTURE

Future developments in society will be strongly influenced by information technology. Computers have gained enormous power by their interaction in global networks. This enables the access to, and exchange of information on a global, and an unprecedented scale.

The present process of business and engineering occurs mainly on a linear scale, one action following the preceding one, cf. Figure 5. The example shows the traditional supply chain, from manufacturing of a product, to transportation, distribution to the retailer and delivery to the customer. The customer has little influence on, and knowledge of the execution and quality of the various steps. This restricts efficiency and makes any modifications or adaptation difficult, once the process has been initiated.

In the future, however, with the introduction of networks, there will be unrestricted exchange of information between all parties involved in the process.

The customer has access to all participants in the supply chain and is being kept informed about the process continuously. Modifications can be made during the execution of the process; in fact the process can be started before all details of the supply chain have been established.

The limitation of the present, traditional networks is that communication of information (text, video, voice etc.) takes place in different networks, cf. Figure 7. Today, we have separated voice, video and data networks, which have evolved and are regulated and tariffed separately.

The new network environment merges the transmission of different types of information as it can support multiple services, cf. figure 8. Such pervasive, converged voice/video/data networks will happen over the next 3 to 5 years and will be challenging to tariff and regulate.

Many applications of merged networks are already emerging in business/commerce (e. g. the banking system), in entertainment and in education. An example is the 'Converged Education System' at Universities.

The above examples are only a glimpse of the many examples which are already taking place on the different networks, both on a global basis (the Internet) and on closed networks (Intranet). The next challenge will be to merge and interlink these networks, providing restricted, or guided access to certain parts of restricted networks (Extranet). The present paper can only give a small, and very limited picture of the many dynamic developments which are presently taking place, and which will happen in the near future.

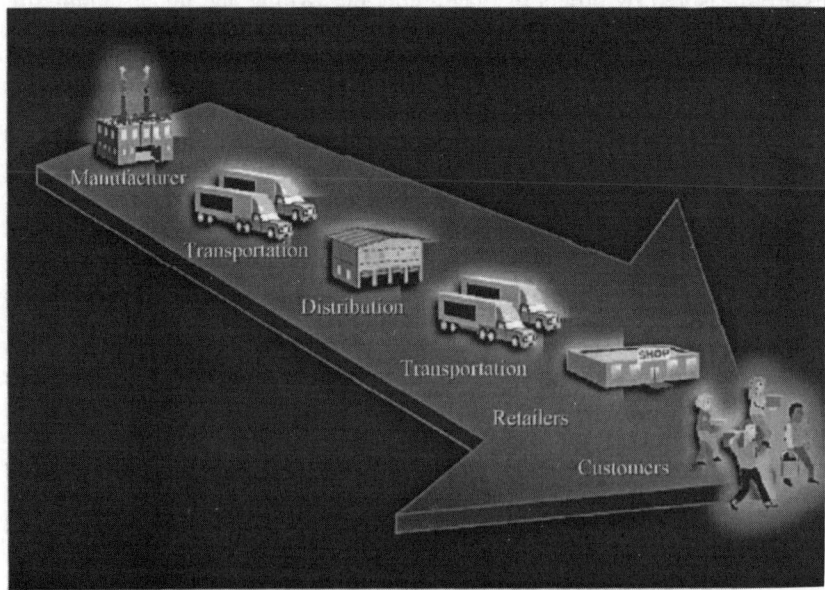

Figure 5. Sequential supply chain of the present economy, courtesy 3 Com.

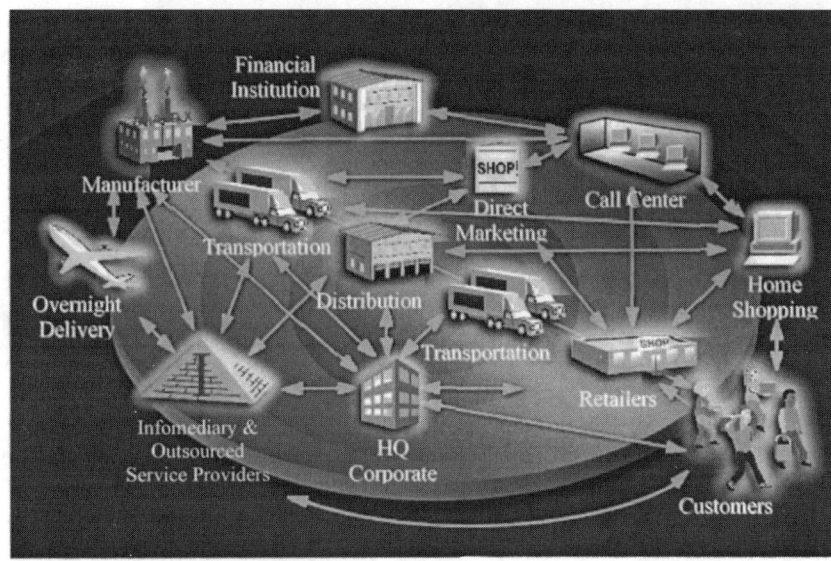

Figure 6. New economy supply chain with network interaction, courtesy 3 Com.

Figure 7. Information transmission in separate networks, courtesy 3 Com.

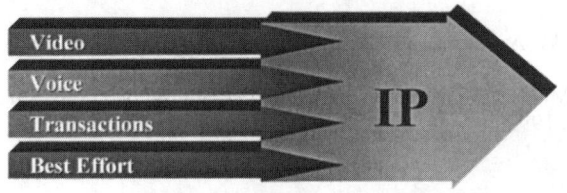

Figure 8. Converged networks which can support multiple services, courtesy 3 Com.

8 ACKNOWLEDGEMENTS

The assistance of Mr. B. Fjelkner of 3 Com in providing information on network structures and the development of network applications is gratefully acknowledged.

Mr. B. Rydell, Swedish Geotechnical Institute was responsible for organising the Workshop on Information Technology and the Geotechnical Profession, at which this paper was presented. His support, patience and encouragement was most valuable.

The Swedish Geotechnical Society and the Swedish Geotechnical Institute, as well as MCIT AB have provided financial support. The advise by A. Massarsch during all phases of this project is appreciated. Prof. J. Hutchinson and A. Lindgren have reviewed the draft manuscript and made several helpful suggestions.

Valuable information on information technology, the history of computer and software was obtained from Encyclopaedia Britannica, Multimedia Edition.

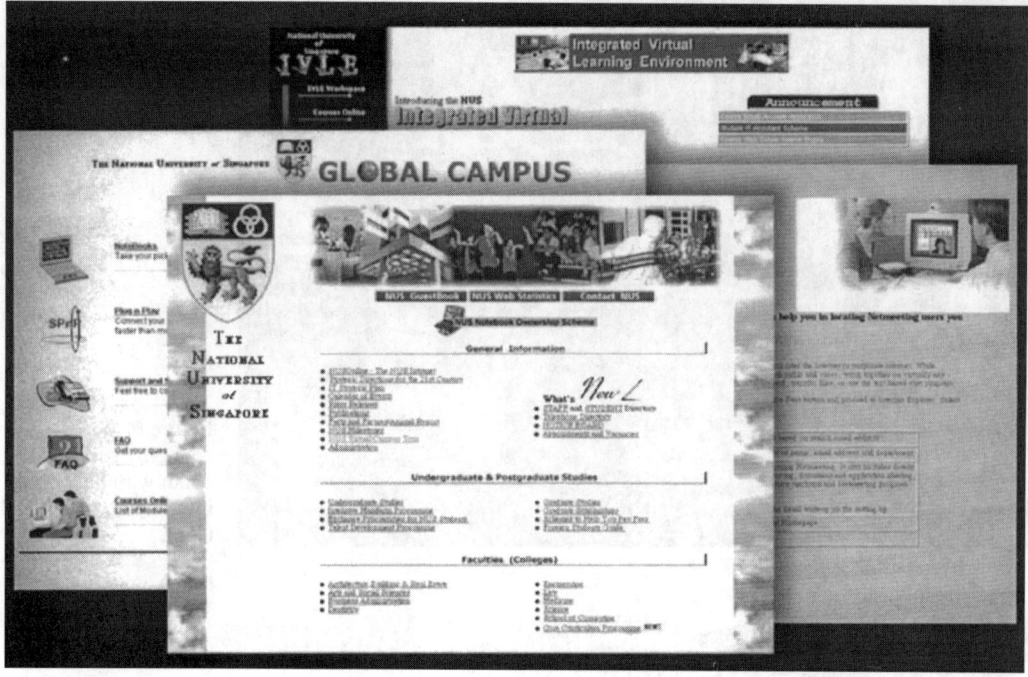

Figure 9. University education using the converged network system

REFERENCES

Bierbier, M. 1982. *The tomb-builders of the Pharaohs*. A Colonade Book, published by the British Museum, 160 p.
Held, G. & R. Sarch. 1997. *Data Communication*. MacGraw-Hill Series on Computer Communications. Encyclopaedia Britannica CD Multi Media Edition. 1999.

Information retrieval and communication

B. RYDELL, A. SALOMONSON & J. LINDGREN
Swedish Geotechnical Institute, Linköping, Sweden

Keywords: IT, internet, information retrieval, communication, databases, literature databases, mailing lists, news groups, information centres, electronic journals, societies, virtual conferences, geotechnology, geotechnics.

ABSTRACT: An important part of all aspects of geotechnical activities, such as research, education, project planning, design and construction, is quick access to relevant information. The recently developed and still emerging tools of Information Technology (IT) offer new, exciting possibilities for efficient access to, and retrieval and exchange of scientific and other technical information. IT tools such as databases, Web sites and e-mail are used extensively today and their present use and future application will be discussed.

This presentation focuses on the importance of IT for information retrieval and communication of scientific and technical information. Access to, and exchange of geotechnical information via the Internet is probably the most important development. Different types of databases, classification of technical information, dissemination and information retrieval via the net are discussed. Furthermore, other methods of communication such as e-mail, mailing lists, news groups and virtual conferences, are described briefly.

The outlook for the near future indicates rapidly extending use of IT in most areas of geotechnical engineering, particularly in education, R&D, design and construction. Knowledge based systems, virtual reality and virtual universities will be essential tools used on a daily basis by students, teachers and scientists, as well as by practising engineers.

The focus of the presentation will be on the use of the Internet for information and communication. The paper includes a list of links to relevant Internet information resources.

1 INTRODUCTION

1.1 *IT creates new forms for working and communication*

During the past decade, Information Technology, or IT, has led to a form of 'information explosion'. In most developed countries, IT is being used by an increasing part of the population in all sectors of society. The Internet has become synonymous with the Information Age. Whatever we call it, the Internet is becoming the primary tool for information and communication, and is limited only by our own imagination. By the year 2000, no less than 1000 million persons around the world are expected to be using the Internet.

With the help of this information and communication technology, we can find new forms for daily work and new meeting places, which can contribute to increasing interest in geotechnology. We can rapidly disseminate new ideas, make new contacts and gain increased access to geotechnical knowledge, literature etc. The younger generation is already growing up with this new technology and it is necessary for all involved to understand the possibilities of IT in order to encourage their interest in geotechnical questions, so important for the development of society. Undoubtedly, the Internet is a very useful tool for achieving this.

1.2 IT so far and in the future...

Extended 'brain power' – in the form of the computer – was 'born' in 1945, when physical power – nuclear weapons – was already great enough to destroy the planet. Neil Armstrong made 'a small step for Man, but a giant leap for Mankind' when he set his foot on the Moon in 1969. When the late president John F Kennedy decided that the US would put a man on the Moon before the end of the decade, this was also the start for information retrieval by computers. Literature database systems such as Lockheed Dialog and ESA-IRS involved databases of different disciplines, from psychology and metallurgy to soil mechanics and foundation engineering. In the geotechnical field, institutes such as the Swedish Geotechnical Institute had systems of their own, but these databases were not accessible from outside.

Since then, the use of IT for information retrieval and communication has increased enormously. Looking back at the trend in IT during recent decades, the focus was on systems from the middle of the 1960s to the beginning of the 1980s. After this, interesting developments were made in the PC era up to about 1994. At present, we are living in the network era with worldwide communication between people. The electronic infrastructure will be further extended and in the future these nets will be as important as the existing infrastructures for water, electricity and transport. During the next millennium, we can expect an increasing importance of the content of information. The winners will be found to be Content Providers, i.e. owners of databases with interesting information necessary for different purposes. Customers will have access to the databases by selected 'filters' and may use agreed information for CAD, calculation, modelling and presentation.

E-mail, intranet and internet can provide a huge increase in communication quality for a company. Examples include, document handling, updated quality manuals on the intranet, internal reference databases, interactive educational programmes, etc. In the future, we will probably be looking at virtual presentations of interactive three-dimensional model programs and projects.

The other side of the coin is increasing IT costs, although communication is the biggest growth area, according to a survey reported in Ground Engineering (May 1999). Almost half of the respondents said that the Internet and e-mail had helped to 'significantly improve the way companies of all sizes do business'. Thus, according to Ground Engineering, 'information technology has become an important part of modern geotechnics over the past few years, with the increase in affordable and user-friendly computing power'. Applications include reporting, data management, project management and quality assurance, as well as accounting and administration.

The survey in Ground Engineering states that about 60 % of geotechnical engineers have access to the Internet, and that 90% of these are able to use it for access to the World Wide Web and e-mail. But what about the issue of free or controlled access? Is free access 'unnecessary and expensive with questionable use'? Does everybody have a professional use for the Internet? Are employees really wasting time searching for information unrelated to work? Questions like these must be answered by each organisation.

2 INFORMATION AND COMMUNICATION NEEDS

The need for geotechnical information is important for the researcher and the consulting engineer, as well as for people involved in the geotechnical process in different ways. Relevant information is also essential for people living in cities or areas where the consequences of geotechnical aspects appear in the form of settlements, landslides or environmental disasters.

Traditionally, the geotechnical engineer's sources of information are informal, e.g. personal communication with colleagues. This form of contact can be widely expanded via the Internet by means of literature retrieval, e-mail, mailing lists, news groups, etc.

Different groups dealing with geotechnical questions in society need different types of information.

Researchers, for example, need to find literature dealing with different areas of their work. Retrieval in databases makes it easier to find whatever is necessary for their research activities. However, there is a paradox: with better tools for finding relevant information, combined with the growth of information, there is a risk of failing to access the necessary information.

Consulting engineers need practically oriented knowledge and experience from cases, as well as information on developments and innovations in geotechnical design, calculations etc. Contractors need relevant geotechnical information for safe and effective production. Case records are important for the exchange of experience from practice.

Customers purchasing geotechnical services or products need knowledge in order to buy relevant services and define quality limits and control activities. Industry, such us producers of equipment, products and chemical substances, needs knowledge on the direction of R&D and the need for geotechnical services for the development of society.

Authorities need information on the possibilities and limitations regarding different types of geotechnical structures for land use planning, as well as their environmental impact and economic consequences.

Finally, the public sometimes needs geotechnical information, for example in areas with landslides or settlement problems, groundwater questions or environmental impact.

3 INFORMATION

3.1 *Information resources*

Today, there is a huge amount of information available worldwide, not least in electronic format. The problem is not mainly the lack of information, but the ability to find the right information required for solving a particular problem. There is a need for guidance through the information jungle, and over the years various tools have been developed by different organisations and publishers.

The following contains an overview of geotechnical information resources. It is important to point out that the presentation does not cover all the possible information resources. For example, the paper does not discuss resources concerning Universities. Instead, it refers to The World-Wide Web Virtual Library of Geotechnical Engineering and its web site at http://geotech.civen.okstate.edu/wwwvl/index.htm, and the section University Geotechnical Engineering Servers.

3.2 Literature databases

A majority of the databases described are so called bibliographic databases containing references to geotechnical literature. The documents are often indexed with keywords and classified according to a classification system. Most of the databases have a summary with a description of the content of the referred document. Appendix 1 presents a list of databases with geotechnical interest. The databases are grouped into two categories: those with a purely geotechnical content and those with a varied quantity of literature references of geotechnical interest. Most of the bases are accessible via the Internet or have information on the Internet showing how to reach them. An access fee is charged for the databases.

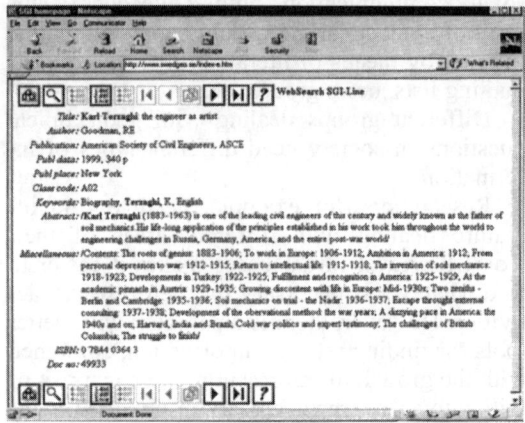

3.3 Information centres/Libraries

To study the contents of the documents retrieved in a bibliographic reference database, it is important to know where to find the documents. Research organisations offer their publications, often in the form of reports, to exchange partners and others. Hard copies or loans of articles in journals, reports and papers from conference proceedings may be ordered from libraries. Publishers provide journals, monographs and conference proceedings. Examples of information centres and publishers will be found in Appendix 2.

3.4 Conference lists

A common way for geotechnicians to keep updated on forthcoming conferences is to obtain bulletins and newsletters from the national geotechnical societies or from lists in journals. During recent years, a large amount of information on forthcoming conferences has become easily accessible on the Internet. For details of where to find the information, see the list of web sites on geotechnical conferences in Appendix 3.

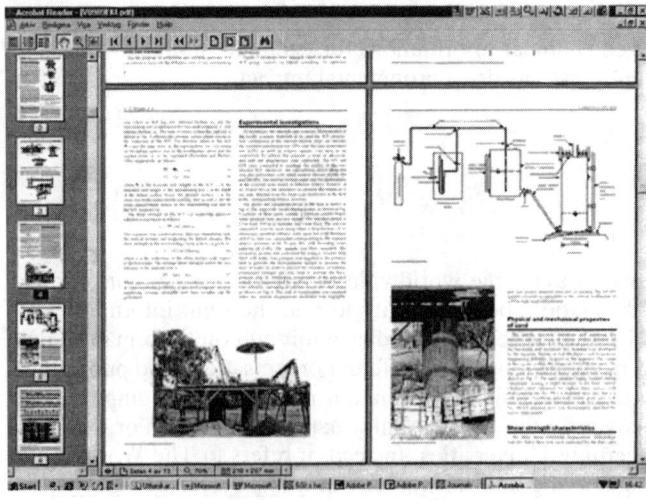

3.5 Journals

Journals are normally distributed by publishers and until a few years ago were only published in printed form. With the birth of the Internet, bibliographic reference information and summaries of articles in journals have begun to be published via the World Wide Web and e-mail. During the 1990s, development of this service has been extended to include the publishing of journals in full text on the Internet, a method which has been adopted by many publishers. Articles in reviewed journals are usually published from half a year up to two years after the manuscript has been sent to the publisher. Publishing via the Internet may shorten this time and thereby result in a more current content. To be able to use this full text service, a subscription has to be taken out. A number of publications containing geotechnical information are listed in Appendix 4.

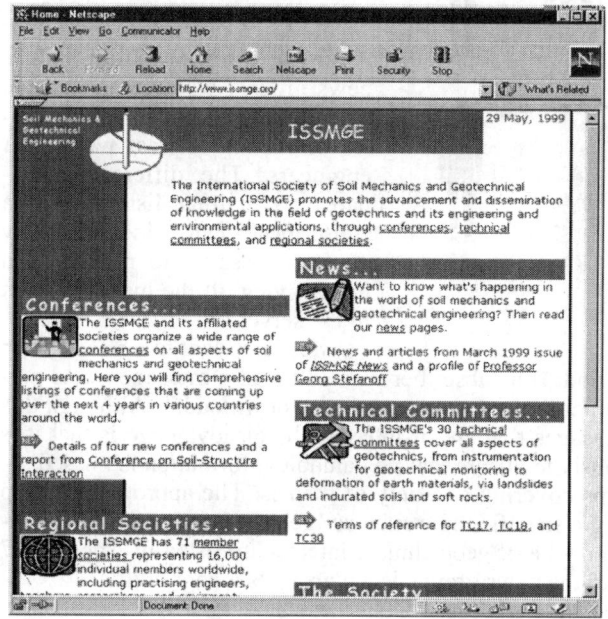

3.6 Societies

Most geotechnicians are members of one or several national or international associations in the geotechnical field. Within these societies, a large number of conferences and courses are held. In the same way, technical committees are established within the societies with the intention of developing different geotechnical areas. Such activities are important for promoting knowledge dissemination and information exchange. Examples of international organisations are given in Appendix 5.

4 COMMUNICATION

Communication possibilities using IT have also developed rapidly in recent years. By using the Internet, it is possible to instantly reach colleagues and information providers, and totally new concepts for establishing working groups can be identified.

4.1 E-mail

E-mail is one of the most popular services on the Internet. According to a survey reported in Ground Engineering (May 1999) some 60% of the geotechnical engineers have access to the Internet and 90% of these are able to use e-mail and the World Wide Web.

4.2 Mailing lists

For communication between a large number of individuals sharing a certain interest or working area, mailing lists can be used.

When a member on a mailing list sends a message, all those on the list receive the information. If a person sends an answer, this goes to every member on the list. Members are registered free of charge on a list server, a computer program that automatically delivers messages to the subscriber of the list. Appendix 6 contains a selection of mailing lists of interest for geotechnical engineers. (Bond, A and Arnould, R, 1998).

Sending messages to/from a mailing list

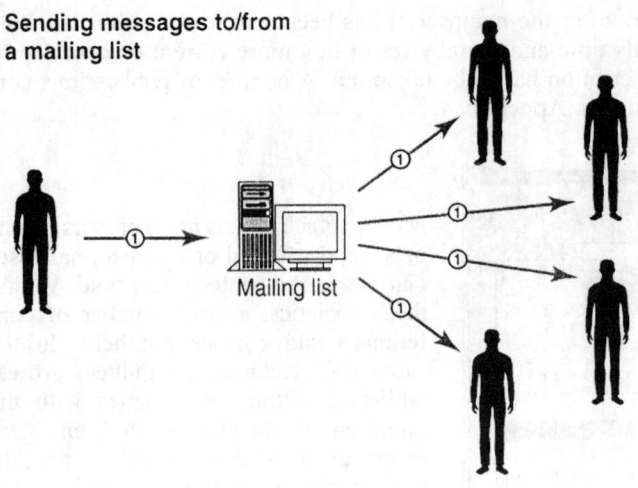

The Appendix provides a brief description of the topics on the mailing lists and addresses for subscribing to the lists.

4.3 News groups

A similar form to mailing lists is news groups, where it is possible to communicate and discuss topics of mutual interest with other engineers. The difference compared to mailing lists is that a subscriber can independently choose messages to read instead of receiving all the messages sent to the server.

Messages sent to a news group will be stored at the server for a limited period. If a subscriber wants to read a message – and perhaps respond to it – it will be necessary to choose the message and then download it. Although this means that a subscriber need only read messages that are of interest, the disadvantage is that it is necessary to follow the discussion continuously to ensure that no valuable information is missed.

There are a large number of news groups covering a variety of themes. The appropriate group can be found by searching the Internet using a specific topic such as 'geo', 'environment' or 'engineering'. Examples of news groups which may be of geotechnical interest are listed in Appendix 7 (Bond & Arnould 1998). One introduction to the news groups is the group 'sci.engr.civil'.

4.4 Virtual conferences

The Internet can also be used to facilitate discussion in virtual conferences. Such conferences have already taken place to some extent. By 'publishing' papers submitted to a conference on the net, it is possible to establish a discussion between interested parties throughout the world. This enables the discussion to continue for a certain period before reaching a conclusion and publishing the results on the web or, if preferred, in paper form.

Another alternative is to establish a discussion prior to a conference in traditional form and thereby focus at the conference on the results, experience and advancements achieved during the web discussion. An example of a virtual conference centre will be found on http://www.mcb.co.uk/confhome.htm.

Even if the Internet offers a number of new forms of information exchange at conferences, there will without doubt continue to be a need for face-to-face meetings.

5 OUTLOOK FOR THE FUTURE OF INFORMATION AND COMMUNICATION

As mentioned above, IT will continue to grow and be used by an increasing number of individuals and organisations. Geotechnical engineers will be able to obtain more and more relevant and useful information. Some information will probably only be available in electronic format.

This will also make it necessary for the engineer to be up-dated on the latest information if he

wants to be competitive in the geotechnical market. A prerequisite for this is that the engineer is active himself in information retrieval necessary for carrying out R&D, making a design or building a geotechnical structure.

Guidance to the relevant information will be provided by the new libraries, working interactively with the engineer. These libraries will form gateways to electronically stored information by publishing literature lists offering access to reference and full text literature databases where the user can search and download the necessary information.

In the near future, we will be seeing new forms or tools for modelling geotechnical structures. With the aid of Virtual Reality, we will be able to work interactively when designing a rock cavern, an offshore plant or an underground railway station. Realistic, moving pictures of the intended structure will be available and will make it possible to visualise the proposed construction in reality.

The engineer will also be able to use artificial intelligence through Knowledge Based Systems, (KBS). Such systems will provide access to a database where established practices or solutions to different problems are compiled.

Information Technology will create an exciting future for mankind in general and for the geotechnical engineer in particular. The possibilities are enormous and we are still at the beginning of the era of advanced application of IT in civil engineering. The limitations still lie in our own imagination and interest in using the new technology.

REFERENCES

Bond, A. (1997). Where to geo on the World Wide Web. Ground Engineering 30(3): 29-30.
Bond, A. (1997). Where to geo on the World Wide Web. Ground Engineering 30(8): 14-15.
Bond, A. & Arnould, R. (1998). Where to geo on the World Wide Web. Ground Engineering 31(8): 16-19.
Information highways and byways. Ground Engineering 32(5): 24.

APPENDICES

Appendix 1. Literature databases

Databases with purely geotechnical contents
Name of database: *GeoFind*
Access: CD-ROM
Database producer: GeoResearch International Inc., 1633 Meadowbrook Road, Ottawa, Ontario, Canada K1B 4 W6
Database description: Bibliographic references to articles published since 1980 in 27 major journals and conference proceedings.

Name of database: *GeoMechanics Abstracts*
Internet access: http://www.orbit.com/english/products/index3.htm
Database producer: Elsevier Science, UK
Database description: Subject coverage includes rock and soil mechanics, properties of geological materials,

engineering geology, hydrology, mining, tunnelling, foundation engineering, rock breakage and excavation, waste disposal, site and laboratory investigations, analysis and design methods.

Name of database: *Geotechnical Abstracts*
Access: CD-ROM
Information: http://www.geotechnical-abstracts.com
Database producer: Research Resources, Inc., USA
Database description: Covers soil mechanics, foundation engineering, rock mechanics and engineering geology.

Name of database: *Geotechnology*
Internet access: http://www.ait.ac.th/clair/centers/geirc/database.htm
Database producer: Geotechnical Engineering International Resources Center (GE-IRC), Asian Institute of Technology, Thailand
Database description: Bibliographic database which contains the major subjects in geotechnical engineering such as soil mechanics, rock mechanics, foundation engineering, engineering geology, earthquake engineering, ground water hydrology and related topics. Database holding is 52,000 references, with an annual increase of 700 references.

Name of database:*SGI-Line*
Internet access: http://www.swedgeo.se/index-e.htm
Database producer: Swedish Geotechnical Institute, Linköping, Sweden
Database description: Bibliographic database containing references to foundation engineering, reinforcement, soil and rock mechanics, site investigations, environmental geotechnics and geology. Most of the documents, books, articles in geotechnical journals, papers in conferences proceedings etc., referred to in the database are available in the SGI Library. Number of references in the database are 50,000, with an yearly increase of 2000.

Databases with varied quantity of literature references of geotechnical interest

Name of database: *Compendex Plus*
Internet access: http://www.eins.org/databases/4.html
Database producer: Engineering Information Inc., New York, USA
Database description: Citations with abstracts to worldwide literature (excluding patents) in engineering and technology. Compendex Plus provides worldwide coverage of approximately 2200 journals and selected government reports and books. In addition to journal literature and publications of engineering societies and organisations, approximately 1000 engineering and technical conferences per year are monitored for indexing. It encompasses all engineering disciplines as well as related fields in science and management.

Name of database: *Enviroline*
Internet access: http://www.eins.org/databases/11.html
Database producer: Congressional Information Service, Inc. Bethesda, MD, USA
Database description: Covers all aspects of environmental sciences. It is an essential source for technical, socio-economic, and policy information concerning the issues affecting the environment. Journals, technical, trade, professional and general periodicals and newspapers, conference proceedings and research reports, are the most important sources for this specialised database. Broad subject categories covered include: Air pollution; Energy; Environment; Land use & misuse; Renewable resources; Transportation; Waste; Water pollution; Noise pollution; Population planning & control; Weather modification & geophysical change; Wildlife.

Name of database: *Geoarchive*
Internet access: http://library.dialog.com/bluesheets/html/bl0058.html
Database producer: Geosystems, Oxon, UK
Database description: Database covering all types of information sources in geoscience, hydroscience, and environmental science. The criteria for inclusion in GeoArchive are that the source should be publicly available and have relevant information content, even if the reference is to a small news item in a magazine. GeoArchive, produced by Geosystems, provides international coverage of over 5000 serials, books from over a 2000 publishers, geological maps, and doctoral dissertations.

Name of database: *Geobase*
Internet access: http://library.dialog.com/bluesheets/html/bl0292.html
Database producer: Elsevier Science
Database description: Bibliographic database covering worldwide research literature in physical and human geography, earth and environmental sciences, ecology, and related disciplines. In addition to providing comprehensive coverage of the core scientific and technical periodicals, Geobase has a unique coverage of non-English language and less readily available publications. Over 2000 journals are fully covered with an additional 3000 having partial coverage. Over 2000 books, monographs, conference proceedings, and reports are also included.

Name of database: *GeoRef*
Internet access: http://library.dialog.com/bluesheets/html/bl0089.html
Database producer: American Geological Institute, GeoRef, 4220 King Street, Alexandria, VA 22302, USA
Database description: Covers worldwide technical literature on geology and geophysics. GeoRef organizes and indexes papers from over 13,000 serials and other publications representative of the interests of the 29 professional geological and earth science societies that are members of the AGI.

Name of database: *IRRD – International Road Research Documentation*
Internet access: http://www.eins.org/databases/43.html
Database producer: OCDE (OECD) Organisation de Cooperation et de Developpement Economique
Database description: The database gathers all literature and information about ongoing research of interest to the road-research community and it is a trilingual (English, French, German) database.

Name of database: *NTIS – National Technical Information Service*
Internet access: http://www.eins.org/databases/6.html
Database producer: US Department of Commerce, Springfield, USA
Database description: Covers scientific, technical, business and economic information contained in publicly-available US government reports. NTIS covers all documents offered for sale by the National Technical Information Service of the US Department of Commerce. It covers research reports, theses, monographs and similar information related to government sponsored research. Subjects covered include: agriculture, astronomy & astrophysics, atmospheric sciences, biological & medical sciences, space science & technology, chemistry, earth sciences & oceanography, electronics, materials, mathematics, mechanical & industrial engineering, civil & marine engineering, military sciences, navigation, communication, detection & countermeasures, nuclear sciences and physics.

Name of database: *Pascal*
Internet access: http://www.eins.org/databases/14.html
Database producer: Institut de l'Information Scientifique et Technique Centre National de la Recherche Scientifique Vandoeuvres-les-Nancy, France
Database description: A multidisciplinary database. Sources include journals, conference proceedings, theses, reports, books and patents. PASCAL covers the core of world scientific and technical literature. In addition there are specialised subfiles in the following areas: Information science, documentation, energy, metallurgy, civil engineering, earth sciences, biotechnology, zoology of invertebrates, agriculture, and tropical medicine.

Appendix 2. Information centres/Libraries

Organisation: *AIT* – Asian Institute of Technology, Geotechnical Engineering International Resources Center (GE-IRC), Asian Institute of Technology, Thailand
Internet access: http://www.ait.ac.th/clair/centers/geirc/

Organisation: *BRE* – Building Research Establishment, UK
Internet access: http://www.bre.co.uk

Organisation: *CIRIA* – Construction Industry Research and Information Association, UK
Internet access: http://www.ciria.org.uk/publications.htm

Organisation: *Geodelft, Netherlands*
Internet access: http://www.geodelft.nl

Organisation: *ICE* – Institution of Civil Engineers, UK
Internet access: http://www.ice.org.uk

Organisation: *LCPC* – Laboratoires Central des Ponts et Chausees, France
Internet access: http://www.lcpc.inrets.fr/LCPC/English/Publications/

Organisation: *NGI* – Norwegian Geotechnical Institute, NGI, Oslo, Norway
Internet access: http://www.ngi.no/english/default.htm

Organisation: *SGI* – Swedish Geotechnical Institute, Linköping, Sweden
Internet access: http://www.swedgeo.se/index-e.htm

Organisation: *TRL* – Transportation Research Laboratory, UK
Internet access: http://www.trl.co.uk

Organisation: *US Army* Corps of Engineers, USA
Internet access: http://lepac1.brodart.com/search/um

Publishers

Name of publisher: American Technical Publishers
Internet access: http://www.ameritech.co.uk

Name of publisher: ASCE Press (American Society of Civil Engineers)
Internet access: http://www.pubs.asce.org

Name of publisher: A.A. Balkema
Internet access: http://www.balkema.nl

Name of publisher: Battelle
Internet access: http://www.battelle.org/publications.stm

Name of publisher: Blackwell
Internet access: http://bookshop.blackwell.co.uk

Name of publisher: CRC Press LLC
Internet access: http://www.crcpress.com

Name of publisher: Elsevier Science
Internet access: http://www.elsevier.nl

Name of publisher: John Wiley & Sons
Internet access: http://www.wiley.co.uk

Name of publisher: Kluwer Academic Publishers
Internet access: http://www.wkap.nl

Name of publisher: McGraw-Hill
Internet access: http://www.books.mcgraw-hill.com

Name of publisher: Prentice-Hall
Internet access: http://www.prenhall.com

Name of publisher: Springer
Internet access: http://www.springer.de

Name of publisher: Technical Standards Service Ltd
Internet access: http://www.techstandards.co.uk

Name of publisher: Thomas Telford
Internet access: http://www.t-telford.co.uk

Appendix 3. Conference lists

Organisation:	Geoforum, Massarsch Constructive IT AB, Stockholm, Sweden
Internet access:	http://www.geoforum.com/contacts/events/index.asp
Organisation:	IAEG - International Association of Engineering Geology
Internet access:	http://www.transport.ntua.gr/IAEG3.html
Organisation:	ISRM - International Society of Rock Mechanics
Internet access:	http://www.lnec.pt/isrm/welcome.html
Organisation:	ISSMGE - International Soil Mechanics for Geotechnical Engineering
Internet access:	http://www.issmge.org/Conferences/conferences.html
Organisation:	SGI - Swedish Geotechnical Institute, Linköping, Sweden
Internet access:	http://www.swedgeo.se/conf.htm
Organisation:	Virtual Library of Geotechnical Engineering
Internet access:	http://geotech.civen.okstate.edu/magazine/confs.htm

Appendix 4. Journals

Name of journal:	Bulletin of Engineering Geology and the Environment
Internet access:	http://link.springer.de
Name of journal:	Canadian Geotechnical Journal
Internet access:	http://www.cisti.nrc.ca/cisti/journals/cjgeoep.html
Name of journal:	Civil Engineering (ASCE)
Internet access:	http://www.pubs.asce.org/ceonline/newce.html
Name of journal:	Civil Engineering. Proceedings of the Institution of Civil Engineers
Internet access:	http://www.t-telford.co.uk/JOL/index.html
Name of journal:	Computers and Geotechnics
Internet access:	http://www.elsevier.nl
Name of journal:	Electronic Journal of Geotechnical Engineering, EJGE
Internet access:	http://geotech.civen.okstate.edu/ejge/index.htm
Name of journal:	Geotechnical Engineering. Proceedings of the Institution of Civil Engineers
Internet access:	http://www.t-telford.co.uk/JOL/index.html
Name of journal:	Geotechnical and Geological Engineering
Internet access:	http://www.wkap.nl/journalhome.htm/0960-3182
Name of journal:	Géotechnique
Internet access:	http://www.t-telford.co.uk/JOL/index.html
Name of journal:	Ground Improvement
Internet access:	http://www.t-telford.co.uk/JOL/index.html
Name of journal:	Hydrogeology Journal
Internet access:	http://link.springer.de
Name of journal:	Internet Geotechnical Engineering Magazine, iGEM
Internet access:	http://geotech.civen.okstate.edu/magazine/index.htm

Name of journal: Journal of Contaminant Hydrology
Internet access: http://www.elsevier.nl/locate/jconhyd or http://www.elsevier.com/locate/jconhyd

Name of journal: Journal of Environmental Engineering (ASCE)
Internet access: http://ojps.aip.org/eeo/

Name of journal: Journal of Geotechnical and Geoenvironmental Engineering (ASCE)
Internet access: http://ojps.aip.org/gto/

Name of journal: Journal of Soil Contamination
Internet access: http://www.crcpress.com/jour/online/sss/

Name of journal: Rock Mechanics and Rock Engineering
Internet access: http://link.springer.de

Name of journal: Soil Dynamics & Earthquake Engineering
Internet access: http://www.elsevier.nl/locate/soildyn/

Name of journal: Soil Science Society of America Journal
Internet access: http://link.springer.de

Appendix 5. Societies

Organisation: IAEG – International Association of Engineering Geology
Internet access: http://www.transport.ntua.gr/IAEG.html
Aims of organisation: To promote and encourage the advancement of Engineering Geology through technological activities and research, to improve teaching and training in Engineering Geology and to collect, evaluate and disseminate the results of engineering geological activities on a worldwide basis

Organisation: IGS – Intenational Geosynthetic Society
Internet access: http://geo.rmc.ca/index.html
Aims of organisation: Dedicated to the scientific and engineering development of geotextiles, geomembranes, related products, and associated technologies

Organisation: ISRM – International Society of Rock Mechanics
Internet access: http://www.lnec.pt/isrm/welcome.html
Aims of organisation: To encourage international collaboration and exchange of ideas and information between rock mechanics practitioners;. to encourage teaching, research, and advancement of knowledge in rock mechanics; to promote high standards of professional practice among rock engineers so that civil, mining and petroleum engineering works might be safer, more economic and less disruptive to the environment.

Organisation: ISSMGE – International Society of Soil Mechanics and Geotechnical Engineering
Internet access: http://www.issmge.org
Aims of organisation: Promotes the advancement and dissemination of knowledge in the field of geotechnics and its engineering and environmental applications, through conferences, technical committees, and regional societies. The ISSMGE has 71 member societies representing 16,000 individual members worldwide, including practising engineers, teachers, researchers, and equipment designers and manufacturers.

Appendix 6. Selected mailing lists of potential interest to geotechnical engineer.

Name of list* *Topics covered*	To subscribe/unsubscribe, send the following message in the body of a (plain text) e-mail:		To post, send you message to...
	Send to...	To subscribe...**/To unsubscribe...	
built-environment* *Construction, engineering, trades, building*	mailbase@mailbase.ac.uk	join built-environment [first] [last] [org]/ leave built-environment stop	built-environment @mailbase.ac.uk
civil-l *Civil engineering research & education*	listserv@listserv.unb.ca	join civil-l [first] [last] [org] *then reply to confirmation request with:* ok/ signoff civil-l	civil-l @listserv.unb.ca
crisp-developers *CRISP finite element program*	mailbase@mailbase.ac.uk	join crisp-developers [first] [last] [org]/ leave crisp-developers stop	crisp-developers @mailbase.ac.uk
crisp-users *CRISP finite element program*	mailbase@mailbase.ac.uk	join crisp-users [first] [last] [org]/ leave crisp-users stop	crisp-users @mailbase.ac.uk
Eevl *Edinburgh engineering virtual library*	mailbase@mailbase.ac.uk	join eevl [first] [last] [org]/ leave eevl stop	Eevl @mailbase.ac.uk
engineering-geotech *All aspects of geotechnical engineering (soil & rock)*	mailbase@mailbase.ac.uk	join engineering-geotech [first] [last] [org]/ leave engineering-geotech stop	engineering-geotech @mailbase.ac.uk
engineering-geotech-mtgs *meetings of teachers of geotechnical subjects*	mailbase@mailbase.ac.uk	join engineering-geotech-mtgs [first] [last] [org]/ leave engineering-geotech-mtgs stop	engineering-geotech-mtgs @mailbase.ac.uk
enveng-l *Environmental engineering*	majordomo@drexel.edu	subscribe enveng-l/ unsubscribe enveng-l	enveng-l@drexel.edu
fea-l *Finite element analysis for mechanical engineers*	listserv@sailor.itis.com	subscribe fea-l/ unsubscribe fea-l	fea-l@sailor.itis.com
Feusers *Finite element analysis*	mailbase@mailbase.ac.uk	join feusers [first] [last] [org]/ leave feusers stop	Feusers @mailbase.ac.uk

* Mailing lists marked * are closed: your request to join will be vetted before being accepted
** In this column, replace [first] with your first name; [last] with your last name; and [org] with the name of your company or organization

Appendix 6. Continued.

Name of list* Topics covered	To subscribe/unsubscribe, send the following message in the body of a (plain text) e-mail:		To post, send you message to...
	Send to...	To subscribe...**/To unsubscribe...	
geo-env *Environmental issues that interface with geology*	mailbase@mailbase.ac.uk	join geo-env [first] [last] [org]/ leave geo-env stop	geo-env @mailbase.ac.uk
Geophysics *Airborne, near-shore, marine, & borehole applications of geophysics*	See instructions on right	*Send blank message to*: join-geophysics @lists.geophysics.co.uk/ *Send blank message to*: remove-geophysics @lists.geophysics.co.uk	geophysics@ lists.geophysics.co.uk
geotech* *Geotechnical earthquake engineering and engineering seismology*	listproc@usc.edu	subscribe geotech [first] [last]/ signoff geotech	geotech@usc.edu
Groundwater *All aspects of groundwater science*	Majordomo @ias.champlain.edu	subscribe groundwater/ unsubscribe groundwater	groundwater@ias.champlain.edu
Hydrology *Australian hydrology*	listserv@eng.monash.edu.au	subscribe hydrology [first] [last]/ unsubscribe hydrology	hydrology@eng.monash.edu.au
soils-l *All aspects of soil science*	listserv@unl.edu	subscribe soils-l/ unsubscribe soils-l	soils-l@unl.edu
tltp-geotechnical *GeotechniCAL computer-aided learning*	mailbase@mailbase.ac.uk	join tltp-geotechnical/ leave tltp-geotechnical stop	tltp-geotechnical @mailbase.ac.uk
unsaturated-soil *Engineering behaviour of unsaturated soils*	mailbase@mailbase.ac.uk	join unsaturated-soil/ leave unsaturated-soil stop	unsaturated-soil @mailbase.ac.uk
yge-l *Young geotechnical engineers' forum*	listproc@itu.edu.tr	subscrbe yge-l [first] [last]/ signoff yge-l *or* unsubscribe yge-l	yge-l@itu.edu.tr

* Mailing lists marked * are closed: your request to join will be vetted before being accepted
** In this column, replace [first] with your first name; [last] with your last name; and [org] with the name of your company or organization

Appendix 7. Examples of News Groups of geotechnical interest

alt.building.construction
alt.building.consulting-specialty
alt.building.contractors
alt.building.engineering
alt.building.environment
alt.building.survey-mapping
alt.construction
alt.geo-software
alt.wastewater

comp.os.geos.misc
comp.os.geos.programmer
cu.civil
cu.engr.general

fj.engr.civil
fj.sci.geo

geo.general
geo.test
gov.us.topic.environment.waste

iu.geosci
iu.geohelp

net.science.geology.misc
ntu.sci.geology

sci.civil.geotechnical

sci.engr
sci.engr.civil
sci.engr.geomechanics
sci.engr.marine.hydrodynamics
sci.engr.mech
sci.engr.mining
sci.environment.waste
sci.geo
sci.geo.earthquakes
sci.geo.fluids
sci.geo.geology
sci.geo.hydrology
sci.geo.mineralogy
sci.geo.oceanography
sci.geo.petroleum
sci.geo.rivers+lakes
scot.environment

ucb.geology
ucd.geology
uiuc.misc.environment
uiuc.org.civil
umich.geo
ut.engineering.general
utexas.geo
utexas.geo.discussion
uw.geoeng
uwo.geog

Information technology applications in geotechnical education and vocational training

D.G. TOLL
University of Durham, UK and Nanyang Technological University, Singapore

ABSTRACT: Information Technology will have an important role to play in geotechnical education and training, and the developments to date clearly demonstrate its potential. A significant number of computer-aided learning packages have already been developed for geotechnical engineering. The materials developed fall into the following categories: Reference materials; Tutorial style activities; Animations or interactive demonstrations; Knowledge-based systems; Simulations; Virtual site visits and Games. Some of these are stand-alone programs that will run on a local computer, whereas increasingly more materials are being mounted on the world wide web. A full listing of nearly forty CAL packages/materials is provided, with website addresses where available.

1 INTRODUCTION

Information technology (IT) is increasingly being used within education and vocational training. Such uses range from 'virtual univeristies' (where course material is made available on the world wide web and tutorial support is provided by email), to computer-aided learning (CAL) packages that can support student learning. This paper focuses on the latter use of IT, since this is the area in which developments relating to geotechnical education and training have been made. However, examples of web based materials are also cited.

A range of computer-aided learning materials have been developed for geotechnical education. Some suites of materials have been developed such as the *GeotechniCAL* project in the UK (Atkinson and Muir-Wood, 1996) and *CATIGE* (Computer Aided Teaching in Geotechnical Engineering) from Australia (Jaska et al, 1996). In addition, there are a number of individual packages.

The types of CAL materials include:
- Reference materials (the equivalent of on-line/electronic text books),
- Tutorial style activities (worksheets, tests and quizzes with supporting materials),
- Animations or interactive demonstrations,
- Knowledge-based systems,
- Simulations (eg finite element analysis),
- Virtual site visits,
- Games.

A brief introduction to each type of material is provided with examples of individual packages being cited within each section. Some packages may combine different types of material. The packages are simply referenced by name (in italics) in the following. A more detailed alphabetical listing of the individual computer-aided learning packages (with references to web sites or publications) is provided later in the paper.

2 REFERENCE MATERIALS

Reference materials are the IT equivalent of text books. They provide support materials for browsing, or for direct reference through indexes or search facilities. They provide text and graphics, and may be enhanced with simple animations. Reference materials have almost exclusively been developed using hypertext approachs. Examples exist of materials developed using Authorware™ (e.g.

ESP), Toolbook™ (e.g. *Tunneling Machinery*), Windows™ Help files (e.g. *Reference*) and Hypertext Markup Language (HTML) for display on the World Wide Web (e.g. *Dam Design* (Fig. 1), *Environmental Geotechnics, FLING, Road Design, Practical Rock Engineering, Slope Design*).

3 TUTORIAL SOFTWARE

Tutorial packages provide worksheets and/or quizzes togther with software to support the different tasks. The best example of tutorial style software is the *Geotutor* package developed at University of West of England, UK as part of the GeotechniCAL suite of materials. This provides a wide range of worksheets together with different types of activity (which comprise simple interactive demonstrations, spreadsheets, reference materials) to assist the student with each task (Fig. 2).

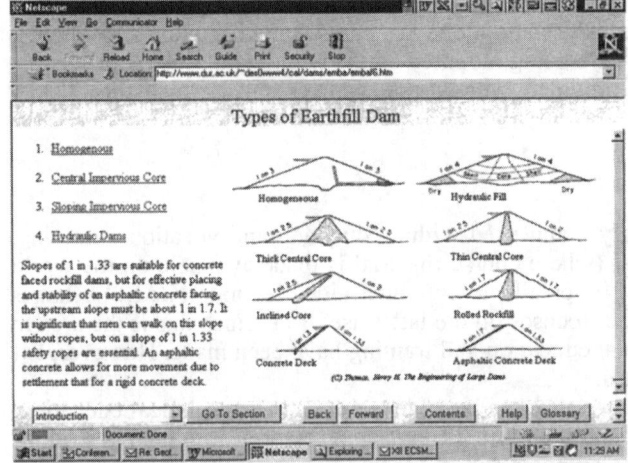

Figure 1. Reference material on Dam Design (University of Durham).

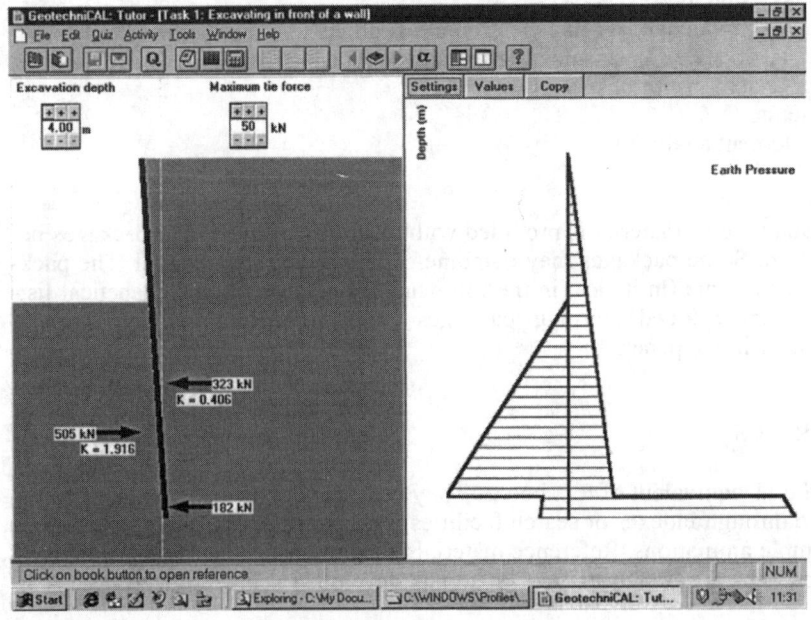

Figure 2. An example of an application within Geotutor (University of West of England).

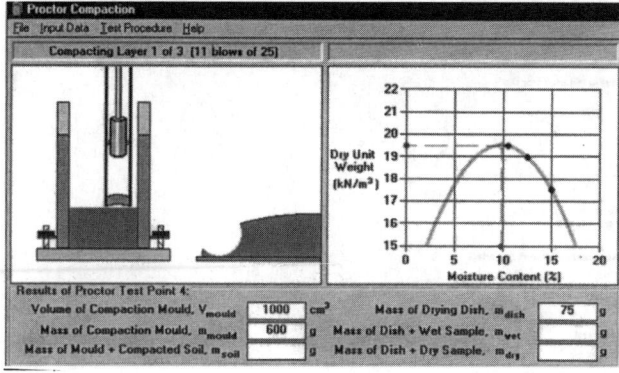

Figure 3. An animated demonstration of Proctor Compaction (University of Adelaide).

4 ANIMATIONS AND INTERACTIVE DEMONSTRATIONS

Animations and interactive demonstrations have been well developed for geotechnical engineering. Such materials have been developed using conventional programming techniques (eg C++) and Virtual Reality Modelling Language (VRML). A number of interactive demonstrations exist for laboratory test equipment, including: Triaxial testing (eg *LABSIM, Triax4W, Virtual Triaxial Test* and the *VRML Triaxial test*); Liquid limit (eg *Characterisation and Classification of Soils, Multimedia Geotechnical Laboratory Testing*), Direct Shear Box (eg *Dsand4W, GeoTutor, Multimedia Geotechnical Laboratory Testing*), Oedometer (*GeoTutor, Virtual Consolidation Test*), Compaction (*ProctorW* (Fig. 3)) and Permeability (*FallingW*). Some allow 'samples' to be prepared and tested. Some have the aim that the student will learn about the test procedures while others intend to demonstrate particular aspects of soil behaviour. There are also applications that demonstrate some simple concepts of geotechnical engineering (e.g. the principle of effective stress (*Effect4W, GeoTutor*), Consolidation (*Consol4W, Consolidation Concept*), stress distribution on a retaining wall (*Retain4W, GeoTutor*).

5 KNOWLEDGE-BASED SYSTEMS

Knowledge-based systems (KBS) have often been identified as suitable for educational purposes; indeed this is frequently the justification for their development. However, KBSs might not be ideal computer aided learning tools, unless they are specifically designed as such. KBS are often referred to as 'expert systems', and by their very nature are often seen as playing an instructional role i.e. they adopt the system control model of directing the user in what to do next. However, KBSs can be developed which do not have these restrictions. An example of such an approach is *ConFound* (Fig. 4) which demonstrates how a KBS (combined with hypertext) can support different learning styles. Rather than the KBS having an instructional role, it can play an advisory role. In this way, students can maintain full control over their own learning.

6 SIMULATIONS

Simulations of soil behaviour allow students to 'play' with different geotechnical applications and to examine the implications of making changes to the geometry of the problem or the soil properties. For simulations (such as finite element analyses) to be successful as computer-aided learning

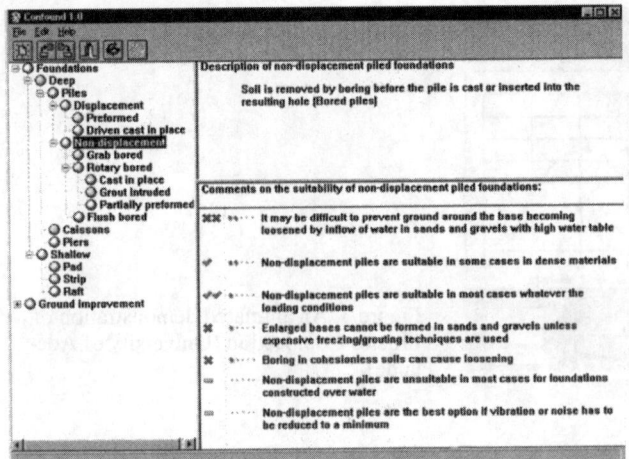

Figure 4. ConFound, a knowledge-based system for conceptual design (University of Durham).

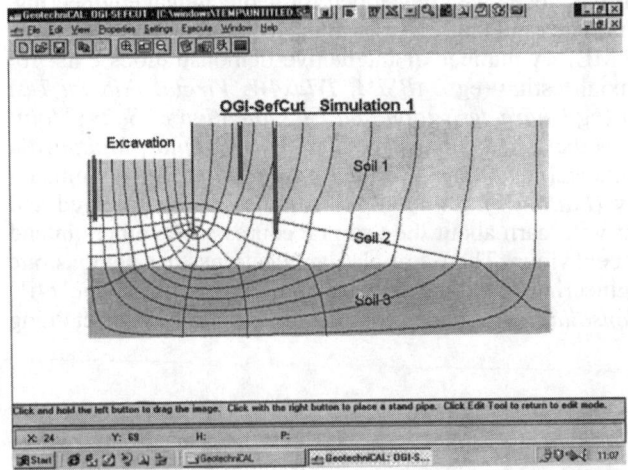

Figure 5. A SefCut simulation (Oxford Geotechnica International).

systems the user interface is all important. A graphical user interface which allows the student to manipulate the problem on-screen is normally required. An example of such a package is *SSI* which allows simulation of foundations or walls. It allows a student to investigate interactions between adjacent footings (for example) and to see how stress conditions and deformations change as the size and spacing of footings is altered. The *Spires* package (Fig. 5) is a similar example for flow problems, allowing the geometry and soil properties to be changed on screen.

7 VIRTUAL SITE VISITS

Site visits to construction sites, or sites of geotechnical and geological interest are an important part of geotechnical education. Unfortunately, time and resources to organise such visits are becoming increasingly scarce. In these circumstances, limited 'real' site visits can be enhanced by, or combined with, 'virtual' site visits. Such materials, normally web-based, provide map and pictorial images of sites of geotechnical interest. Good examples are *CIVCAL* (Fig. 6) and *Coastal Development in the Moray Firth*. GEOMECA also provides many images of construction or of geotechnical and geological interest, but these are not provided with a commentary.

Information technology applications in geotechnical education and vocational training 103

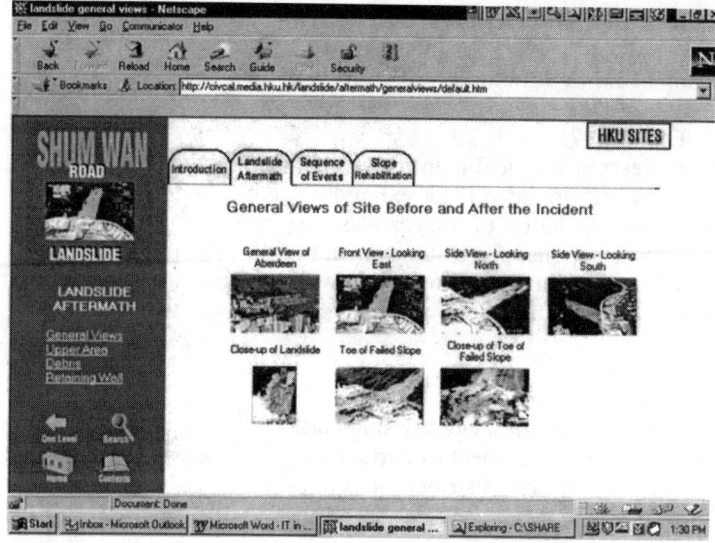

Figure 6. A virtual site visit to a landslide site (University of Hong Kong).

Figure 7. The starting point for the Site Investigation game (South Bank and Portsmouth Universities).

8 GAMES

Games can provide a powerful learning aid by making learning 'fun'. Students are often familiar with video games in their non-study time and may respond well to such initiatives for study purposes. However the resources required to develop games for geotechnical engineering cannot usually compete with the resources used to develop commercial games software. Therefore such attempts may appear to be poor quality in comparison and can deter students from their use. Nevertheless, a successful example of a geotechnical game is the *Site Investigation* package (Fig. 7). The use of multi-media and humour make this an excellent learning package that students enjoy using (Thompson & Toll, 1997a,b).

9 LISTING OF COMPUTER-AIDED LEARNING MATERIALS

CAVES (Computer Aided Visualisation of Earth Structure)
Herriot-Watt University, UK
Reference: Paul and Balfour (1990), Paul (1997)
CAVES was designed to help students learn geological map interpretation. It produced images of 3D geological structures which could be manipulated. The examples were based on a standard textbook. The advantages were that the manipulation of images was very appealing, it aided visualisation of 3D structure and it had a good response from students and improved student performance. However, it ran as a stand-alone program for RM Nimbus and was too hardware specific and therefore difficult to maintain or extend. It lasted only five years and is no longer in use.

Characterisation and Classification of Soils
Instituto Militar de Engenharia, Rio de Janeiro, Brazil
This CD (in Portuguese) describes the procedures for characterising and classifying soils according to the HRB and USCS schemes. It describes measurement of particle density, water content, liquid limit, plastic limit, shrinkage limit and particle size distribution. Video clips of the tests are included.

CIVCAL
University of Hong Kong, Hong Kong University of Science and Technology, Hong Kong Polytechnic University, The City University of Hong Kong
Web Site: http://civcal.media.hku.hk/
CIVCAL comprises an electronic data base of mainly pictorial material surrounded by an access and manipulation shell and a set of complementary interactive learning and problem solving tools. The authors suggest that, in many ways, an actual site visit may be adequately replaced and, in fact, in some respects enhanced through a virtual site visit that makes use of computer and multimedia technology. Large amounts of visual, audio and narrative material were obtained from recently constructed projects in Hong Kong and, in some cases, from projects actually under construction. In addition, computer-based multimedia materials have been developed for use in conjunction with lectures, laboratory sessions, tutorials and for independent study to assist students in developing competence in solving engineering problems.

Class4W (CATIGE)
University of Adelaide, Australia
Web site: http://www.eng.adelaide.edu.au/CATIGE/main.html
Reference: Jaska et al (1996)
Class4W guides students through the process of identifying and classifying soils using the Unified Soil Classification System (USCS). Class4W uses CATIGE's six hypothetical soils and allows the user to choose various laboratory tests and field identification techniques to identify the soils. The results of liquid limit tests and sieve analyses can be plotted to assist the student in classifying the soils. In order to make the process realistic, the student is given a budget and each laboratory test is charged against this budget. The student is asked to suggest the USCS symbol for the selected soil.

Coastal Development in the Moray Firth (MARBLE)
Heriot-Watt University, UK
Web site: http://www.civ.hw.ac.uk/online.html
Reference: Paul (1997)
This web-based courseware is designed to support practical classes on environmental change in the Moray Firth, Scotland. It provides an introduction to the Moray coast, prior to a desk study, and includes a virtual field trip to Ardersier. It uses topological maps, geological maps, aerial photos and ground photos.

ConFound (GeotechniCAL)
University of Durham, UK
Web site: http://geocal.uwe.ac.uk/
References: Toll & Barr (1996; 1998)
ConFound is a knowledge-based system (KBS) for the conceptual design of foundations. The student can enter project specific information under 3 categories: information about the structure (e.g. loading, tolerance to movement); information about the site (e.g. past uses, topography); information about the ground (e.g. soil/rock conditions, test results). The system then displays a hierarchical list of foundation types. If the student selects a foundation type then a brief description of that type is provided. Also provided is a list of comments about the suitability of that type for the conditions specified by the student, ranked in order of confidence level. These comments come from the knowledge base which comprises rules, each with a series of comments (and link to reference material in a help file) and values quantifying the suitability and degree of confidence in the rule. Each foundation type can have any number of rules associated with it. At any stage the student can find more detailed information about any point raised by the KBS by clicking on a comment which displays the appropriate page of the reference materials.

Consol4W (CATIGE)
University of Adelaide, Australia
Web site: http://www.eng.adelaide.edu.au/CATIGE/main.html
Reference: Jaska et al (1996)
Consol4W provides an introduction to the processes that occur during one-dimensional consolidation. Consol4W allows the student to choose one of the standard soils, one- or two-way drainage, the thickness of the consolidating layer, the stress increment, and the time interval between results. Consol4W displays the consolidating layer as well as graphs of excess porewater pressure vs. depth of the layer, and the change in layer thickness vs. time.

Consolidation Concept
University of Arizona, USA
Web site: http://www.u.arizona.edu/~budhu/courseware.html
The Consolidation Concept is an interactive simulation of the process of consolidation of fine-grained soils. Students interact with the software to get an understanding of the important aspects of the time dependent settlement of soils.

Dam Design
University of Durham
Website: http://www.dur.ac.uk/~des0www4/cal/dams/fron/startup.htm
Reference: Graham (1997)
These web pages for dam design provide comprehensive information on concrete and embankment dams. There are sections on loading, site investgation, geology, hydrogeology, foundations, spillways, and construction of dams. There are also some limited self-assessment questions and worked examples, a glossary and reference lists.

Dams4W (CATIGE)
University of Adelaide, Australia
Web site: http://www.eng.adelaide.edu.au/CATIGE/main.html
Reference: Jaska et al (1996)
Dams4W illustrates the two-dimensional flow beneath a dam, sheet-pile, or other user-defined retaining structure. The user can specify boundary hydraulic heads, soil permeability and the geometry of the retaining structure. By choosing a location within the flow field, Dams4W plots a flow velocity or equipotential vector. Performing this at several locations within the flow field, the user may generate a flow net consisting of flow lines and equipotential lines. The emphasis of Dams4W is not to produce a flow net but to facilitate the task and to involve the student in the process.

DSand4W (CATIGE)
University of Adelaide, Australia
Web site: http://www.eng.adelaide.edu.au/CATIGE/main.html
Reference: Jaska et al (1996)
DSand4W is a graphical representation of the direct shear test performed on specimens of sand. The student may select either dry sand, or a saturated sand with water pressure, tested in a loose, medium, or dense state. After specifying the hanger load, DSand4W animates the test apparatus and plots the result on a shear stress vs. displacement and a normal stress vs. peak shear stress graph. The user is then able to perform additional tests with different normal stresses, after which, the student can estimate the angle of friction, ϕ.

Effect4W (CATIGE)
University of Adelaide, Australia
Web site: http://www.eng.adelaide.edu.au/CATIGE/main.html
Reference: Jaska et al (1996)
Effect4W seeks to reinforce the understanding of vertical effective stresses. Up to four separate soil layers may be entered with different void ratios, bulk unit weights, moisture contents, and specific gravities. Effect4W plots the total and effective stresses and the porewater pressure as a function of depth, and allows the user to view the effect of varying the depth of the water table.

Environmental Geotechnology
Bolton Institute, UK
Website: http://www.technology.bolton.ac.uk/civils/mscenvgeo/
This is a set of web-based courseware to support a MSc in Environmental Technology at Bolton Institute. Existing materials relate to contaminated land. They cover history and political initiatives; soil assessment; water assessment and reclamation, including innovative treatment methods. Materials on ground investigation are under development.

ESP (Effective Stress Program)
Heriot-Watt University, UK
Reference: Oliver and Oliphant (1999)
ESP provides support to students who are learning the principle of effective stress and its application. It adopts a linear 'supervised' approach where the student follows a set sequence of topics: Total vertical stress, Pore water pressure and Effective stress. Within each topic the student is required to work through a Theory section, followed by Worked Examples, and finally a Test section (to test the student's understanding). The test sections provide a high degree of interactivity. The program reacts to students' answers and provides informative feedback when mistakes are made, advising how an error can be corrected.

Expansiv (CATIGE)
University of Adelaide, Australia
Web site: http://www.eng.adelaide.edu.au/CATIGE/main.html
Reference: Jaska et al (1996)
Expansiv allows the student to enter a soil profile characterised by the number of soil layers, their thicknesses and instability indices and the soil suction profile. Expansiv calculates the amount of surface heave associated with the soil profile, and displays the distortion of a residential dwelling as a function of this heave, as well as a description of the process that is occurring. In addition, external factors such as seasonal effects; leaking services; poor stormwater drainage and tree effects can also be examined.

FallingW (CATIGE)
University of Adelaide, Australia
Web site: http://www.eng.adelaide.edu.au/CATIGE/main.html
Reference: Jaska et al (1996)

FallingW provides the student with an introduction to the measurement of the permeability of soils by means of the falling head test. After choosing a soil, air pressure and time interval, FallingW animates the test, and the head of the water in the standpipe is plotted against time. The student is able to start and stop a timer, thereby enabling values to be recorded throughout the test. After completing the test the student is able to evaluate the permeability of the soil.

FLING (Flexible Learning in Geotechnics)
Heriot-Watt University, UK
Web site: http://www.civ.hw.ac.uk/online.html
Reference: Paul (1997)
FLING has been developed to support a MSc module in Environmental Geotechnics provided jointly by Heriot-Watt and Glasgow Universities. It provides lectures notes and illustrations, lists of references (with abstracts), case studies, frequently asked questions and self-assessment tests. It is intended to reinforce limited class contact time. Materials include the investigation of contaminated land. A unit on landfill is under development.

GEOMECA
Ecole Centrale Paris, France
Web site: http://geomeca.ecp.fr/
GeoMeca is the web site of KSO (Knowledge Synthesis Organisation). Their aim is to provide a better understanding of the properties of the in situ soils for tunnels, foundations, dams and roads. The main feature of the website is an extensive database of photographs of all types of construction or views of geotechnical interest. These include sites from France, Belgium and Portugal. There are also links to multimedia presentations (in French) on soil behaviour and video clips.

GeoTutor (GeotechniCAL)
University of West of England, UK
Web site: http://geocal.uwe.ac.uk/
References: Davison & Porritt (1999)
The GeoTutor program provides activities, spreadsheets and self-assessment quizzes, linked to the hypertext reference information It is provided with paper-based workbooks (available as Microsoft Word files) which together with GeoTutor introduce the principles of geotechnical engineering. For each task in the workbook, there is a corresponding part of GeoTutor to be selected. GeoTutor enables students to explore some of the important concepts by manipulating simple models and observing the effect. It contains some 43 activities, all designed to support students working through the workbook tasks. The activity settings (including soil parameters and spreadsheet formulae), quiz questions and guided tours are stored in simple text files which can all be edited by the tutor.

Heave (CATIGE)
University of Adelaide, Australia
Web site: http://www.eng.adelaide.edu.au/CATIGE/main.html
Reference: Jaska et al (1996)
Heave calculates the surface heave, y_s, and the design heave, y_m, of an expansive soil profile in accordance with Australian Standard AS 2870. In addition, Heave calculates the influence of trees on y_s.

LABSIM (GeotechniCAL)
University of Glasgow, UK
Web site: http://geocal.uwe.ac.uk/
LABSIM provides a computer simulation of the triaxial test. The principal objective and mode of use of LabSim is to provide students with the opportunity to carry out triaxial tests on-screen, when physical and personnel resources are insufficient to allow individual experimentation in the laboratory. The emphasis is on understanding soil behaviour and, only secondarily, training in test pro-

cedures. A companion program (Configur) allows tutors to modify the LabSim simulation, to reflect their own requirements for technical content, soil types, quiz questions and learning styles. The main screen of the LabSim simulation depicts a schematic of the triaxial cell, complete with cell pressure gauge, pore water pressure gauge, load cells, drainage ports, clock etc. During the consolidation phase, three plots are displayed (versus time): mean effective stress, pore water pressure and volume change. During the loading phase, the on-screen plots are: pore water pressure response, axial stress-strain response, stress paths, and volume change.

MECSOLOS (REESC)
Federal University of Santa Caterina, Brazil
Web site: http://reesc.ctclab.ufsc.br
Reference: Ferreira (1998)
The MECSOLOS software makes use of text (in Portuguese), images, graphical illustrations and animation to help students to learn the concept of the stress distribution in soils. This forms one application of the REESC Project (Reengineering of the Engineering Education in Santa Catarina) which involved seven universities of the state of Santa Caterina.

Mohr4W (CATIGE)
University of Adelaide, Australia
Web site: http://www.eng.adelaide.edu.au/CATIGE/main.html
Reference: Jaska et al (1996)
Mohr4W demonstrates two-dimensional stress transformation by means of the Mohr circle. An element of soil is displayed with user defined values of horizontal and vertical stresses, and as the user rotates the element, Mohr4W plots a vector representation of the normal and shear stresses, and plots the Mohr circle.

Multimedia Geotechnical Laboratory Testing
University of Idaho, USA
Reference: Sharma and Hardcastle (1999a,b)
This multimedia CD is intended to demonstrate geotechnical laboratory test procedures. Each module contains reference material, modelling and simulation using 3D graphics and video and an interactive tutorial and quiz to test the students' understanding. It has been developed using Toolbook™. Modules will include: Water content, weight-volume relationships; Visual soil classification; Atterberg limits; Grain size distribution; Compaction tests; Field density testing; Permeability tests; Consolidation testing; Direct shear test and Triaxial tests.

Multimedia Soil Laboratory
University of Portsmouth, UK
Reference: Alani, A. and Barnes, R. (1999)
This presently exists only as a pilot version containing one test but is actively under development.

Practical Rock Engineering
Rocscience Inc., Toronto, Canada
Web site: http://www.rocscience.com/Hoekcorner.htm
This web site contains the complete text (in pdf format) of a set of notes on Practical Rock Engineering by Evert Hoek. The 16 chapters include: Development of rock engineering; When is a rock engineering design acceptable; Rock mass classification; Shear strength of discontinuities; Structurally controlled instability in tunnels; The Rio Grande project - Argentina; A slope stability problem in Hong Kong; Factor of safety and probability of failure; Analysis of rockfall hazards; In situ and induced stresses; Rock mass properties; Tunnels in weak rock; Large Powerhouse caverns in weak rock; Rockbolts and cables; Shotcrete support; Blasting damage in rock.

ProctorW (CATIGE)
University of Adelaide, Australia
Web site: http://www.eng.adelaide.edu.au/CATIGE/main.html
Reference: Jaska et al (1996)
Proctor4W demonstrates the Proctor and modified Proctor compaction tests. The student may choose one of CATIGE's six hypothetical soils and the type of Proctor test. The process is demonstrated by using an animated graphics screen and, if desired, sound. Proctor4W guides the student through the compaction test procedure and plots the results on a standard compaction graph. The student is able to add or remove moisture and repeat the test, enabling several compaction points to be determined. Having done this, the student is then asked to estimate the optimum moisture content and the maximum dry unit weight of the soil.

Reference (GeotechniCAL)
University of West of England, UK
Web site: http://geocal.uwe.ac.uk/
References: Davison & Porritt (1999)
Reference contains a hypertext geotechnical reference manual. It is a collection of Windows™ Help files, one for each subject: Basic mechanics, Soil mechanics, Groundwater, Foundations, Retaining walls and Slope stability. Each file contains a structure of short pages, giving an overview of the subject, leading to levels of increasing detail, plus summaries of case studies and references to journals and texts. Definitions of terms are provided by pop-up entries from the glossary. Many of the diagrams and symbols also have hot-spots with pop-up labels and definitions.

Retain4W (CATIGE)
University of Adelaide, Australia
Web site: http://www.eng.adelaide.edu.au/CATIGE/main.html
Reference: Jaska et al (1996)
Retain4W is a sheet pile retaining wall analysis program. It demonstrates the analysis of cantilever sheet pile retaining walls based on Rankine earth pressure theory. Retain4W allows the student to input different soil properties and water tables on both the active and passive sides of the wall, and calculates the sliding forces, overturning moments, and the factors of safety against sliding and overturning. If the factors of safety are less than one, Retain4W animates the wall and displays its collapse. The animation is dependent on which mode of failure occurs.

Road Design
University of Durham, UK
Website: http://www.dur.ac.uk/~des0www4/cal/roads/framettl.html
Reference: Wilkinson (1997)
These web pages contain reference materials concerning road design, based mainly on the UK Department of Transport's Design Manual for Roads and Bridges. There are sections on History; Traffic analysis; Site Investigation; Earthworks and Pavement Design.

Site Investigation (GeotechniCAL)
South Bank University & University of Portsmouth, UK
Web site: http://geocal.uwe.ac.uk/
References: Moran et al (1997)
The package comprises a Site Investigation game, supported by a series of tutorial modules. The game enables students to encounter, through images, animation, video and audio, the challenges of the real life site investigation. The tutoring modules are designed to enable students to learn about the basic constituents of a site investigation, to see the range of information available and its usage.

Slope Design
University of Durham, UK

Web site: http://www.dur.ac.uk/~des0www4/cal/slopes/index.html
Reference: Connolly (1997)
These web pages on contain reference materials on slope stability. There are sections on Introduction to slopes; Introduction to slope instability; Slope stability analysis; Remedial and corrective measures for failing slopes and Case studies. There are also some limited example slope stability problems (without solutions).

Spires (GeotechniCAL)
Oxford Geotechnica International, UK
Web site: http://geocal.uwe.ac.uk/
This package contains three programs: SefCut, SefDam and SefWeir. They enable the student to modify the parameters in standard seepage flow problems, and calculate the two-dimensional flow nets. *Sefcut* models flow into an excavation in layered soil supported by a sheet pile wall. Excavation size, sheet pile depth, layer thickness and permeability can all be altered. *SefDam* demonstrates flow through a simple earth dam with a core of different permeability. The side slopes, crest width, location and width of the core, upstream and downstream water levels can all be altered, as well as the permeabilities of the dam and core materials. *SefWeir* models flow below an impermeable structure, with the possibility of including a cut-off. The location and depth of the cut-off can be altered.

SSI (GeotechniCAL)
South Bank University and University of Surrey, UK
Web site: http://geocal.uwe.ac.uk/
SSI stands for Soil-Structure Interaction. The package allows students to study structural elements embedded in, or resting on, the ground. Examples include foundations and retaining walls. In each of these cases the response of the system depends on the stiffnesses of both the soil and the structure as well as the loads which are applied. Increasingly engineers in industry are using such as the finite element method when Soil-Structure Interaction effects are important. The underlying method used by the program is based on finite elements. However, the detail of the process of constructing the mesh is completely hidden from students. This is deliberate, as it authors of the package believe that the basic skill of interpreting the results does not depend on detailed knowledge of finite element theory or how to create a mesh.

Triax4W (CATIGE)
University of Adelaide, Australia
Web site: http://www.eng.adelaide.edu.au/CATIGE/main.html
Reference: Jaska et al (1996)
Triax4W simulates the triaxial testing of soils. All six of CATIGE's soils can be tested, using drained or undrained conditions. The axial stress and cell pressure can be increased or decreased during the test, and the drainage valve can be opened or closed at any time. An axial stress-axial strain graph and a p, p', q stress path can be plotted. Porewater pressures are measured and displayed throughout the test.

Tunneling Machinery
Ecole Centrale Paris, France
Web site: http://geomeca.ecp.fr/
This multimedia CD on Tunneling Machinery (in French) was developed by J. Biarez and S. Taibi at Ecole Centrale Paris. It has been implemented in Toolbook™ and has extensive images of the different types of tunneling machinery and case studies relating to tunneling. It can assist with the choice of tunneling methods based on information on the types of rock or soil and hydrogeology.

Virtual Consolidation Test
University of Arizona, USA

Web site: http://www.u.arizona.edu/~budhu/courseware.html
Reference: Budhu, M. (1999a, b)
The Virtual Consolidation test package provides an interactive simulation of the oedometer test. It is not intended to eliminate the actual test but to complement hands-on experience; to extend the range and convenience of testing; to test a student's prior knowledge; to guide the student through the testing; to allow the student to prepare and interpret the test results, and to evaluate the student's understanding of the calculations involved and the interpretation of the test results. Throughout the exercises, the student can monitor her/his performance and seek on-line help.

Virtual Triaxial Test
University of Arizona, USA
Web site: http://www.u.arizona.edu/~budhu/courseware.html
Reference: Budhu, M. (1999a, b)
The Virtual Triaxial test packages provides an interactive simulation of the triaxial test. It has the same aims as the Virtual Consolidation Tests (see above).

VRML Triaxial Test
University of West of England, UK
Web site: http://geocal.uwe.ac.uk/
References: Davison & Porritt (1999)
This is a demonstration of the use of VRML (Virtual Reality Modelling Language) to provide a 3-D simulation of the triaxial test. It allows the student to place the sample in the cell, to consolidate and to apply axial loading. It is still a simple prototype, but demonstrates the potential for virtual reality as a teaching/learning tool.

10 CONCLUSIONS

A range of high-quality computer-aided learning materials have been developed for Geotechnical Engineering. These comprise: Reference materials; Tutorial style activities; Animations or interactive demonstrations; Knowledge-based systems; Simulations (eg finite element analysis); Virtual site visits and Games. Many packages are stand-alone programs which will run on a local computer. However, some materials (particulaly reference materials) are available through the world wide web. Such materials are able to provide support for learning in areas that are difficult to cover with traditional teaching and they can provide a useful compliment to other teaching methods.

REFERENCES

Alani, M. & Barnes, R. 1999. *A multimedia soil mechanics laboratory software development for teaching and learning purposes,* International Conference in Engineering Education ICEE 99, Ostrava, Czech Republic, August 1999, Paper Reference Number 367.

Atkinson, J.H. & Muir-Wood, D. 1996. *Byte the hand that needs,* Ground Engineering, 29, 7 (Sept.), pp 34-35.

Budhu, M. 1999a. *A Simulated Soils Laboratory Test,* Proc. Int. Conf. on Simulation and Multimedia in Engineering Education, ICSEE'99 (eds Tharp, H. & Huelsman, L.), pp. 3-6.

Budhu, M. 1999b. *Multimedia Geotechnical Laboratory Test Courseware,* 1999 ASCE Annual Conference & Exposition, Charlotte, NC, June 20-23, 1999.

Connolly, H. 1997. *World Wide Web Pages for Slope Design*, MEng final year project report, School of Engineering: University of Durham, 43 pp.

Davison, L. And Porritt, N. 1999. *Using computers to teach,* Proceedings of the Institution of Civil Engineers, Civil Enginering, Volume 132 (Feb.), pp. 24-30.

Ferreira, R.S. 1998. *Learning Stress Distribution in Soils Using a Digital Multimedia Tool*, International Congress of Engineering Education, Rio de Janeiro, Paper 466 (available at http://www.ctc.puc-rio.br/icee-98/).

Graham, A. 1997. *The Development of World Wide Web Pages for Dam Design*, MEng final year project report, School of Engineering: University of Durham, 50 pp.

Jaska, M.B., Kaggwa, W.S. and Gamble, S.K. 1996. *CATIGE for Windows – A Computer Aided Teaching Suite for Geotechnical Engineering*, Proc. 7th Aust. NZ Conf. on Geomechanics, Adelaide, pp. 976-980.

Moran, J. A., Langdon, N. J. and Giles, D.P. 1997. *Can site investigation be taught?* Proceedings of the Institution of Civil Engineers, Volume 120, Issue 3, pp. 111-118.

Oliver, A.W. & Oliphant, J. 1999. *A Computer-Aided Learning Program for Teaching Effective Stress to Undergraduates,* Geotechnical and Geological Engineering, Vol. 17.

Paul, M.A. 1997. TALiSMAN *Specialist Seminar on Copyright and the Web,* Moray House Institute of Education, Edinburgh, 23 June 1997, http://www.talisman.hw.ac.uk/.

Paul, M.A. & Balfour, J.A.D. 1990. *A computer assisted approach to the teaching of geological map interpretation to undergraduate Civil Engineers*, Proceedings of the Institution of Civil Engineers. Part 1. Vol. 88, pp. 367-80.

Sharma, S. & Hardcastle, J.H. 1999a. *Using Multimedia to Teach Geotechnical Engineering Laboratory Procedures*, 34th Engineering Geology & Geotechnical Engineering Symposium, Utah State University, Logan, UT, April, 1999.

Sharma, S. & Hardcastle, J.H. 1999b. *A Multimedia Approach for Teaching Geotechnical Engineering Laboratory Testing,* Proc. 11th Panamerican Conference on Soil Mechanics & Geot. Engrg., Iguasu Falls, Brasil, August, 1999.

Thompson, L.A. & Toll, D.G. 1997a. *Evaluating Students' and Tutors' Perceptions of Computer Assisted Learning Materials: A Case Study,* Innovations in Education and Training International, Vol. 34, No. 4, pp 272-280.

Thompson, L.A. & Toll, D.G. 1997b. *I like this but..... Student Evaluations of Computer Assisted Learning Materials,* Habitat, Issue 3, pp 17-19 (also available at http://ctiweb.cf.ac.uk/HABITAT/HABITAT3/).

Toll, D.G. & Barr, R.J. 1996. *Computer-aided Learning for Geotechnical Engineering,* Deliberations on Teaching and Learning in Higher Education, JISC Electronic Libraries Programme, http://www.lgu.ac.uk/iem/engineering/eng_comp.html.

Toll, D.G. & Barr, R.J. 1998. *A Computer-aided Learning System for the Design of Foundations,* Advances in Engineering Software, Vol. 29, No. 3, pp 637-643.

Wilkinson, D. 1997. *WWW Pages for Road Design*, MEng final year project report, School of Engineering: University of Durham, 42 pp.

DESSYS (Geotechnical Design Decision System)

M. VAN VEGHEL
Eindhoven University of Technology, Department of Architecture, Building and Planning, Design Systems Group, Eindhoven, Netherlands and Fugro Ingenieursbureau B.V., Leidschendam, Netherlands

Keywords: Geotechnical design decision support system, VR-DIS, expert systems, multidisciplinary design, collaborative design

ABSTRACT: Design is the work of a multi disciplinary design team that in the near future probably will be carried out in virtual design studios. We believe that the next generation of design systems should support the ability to communicate, collaborate and coordinate design decisions with clients, users, legislators, advisors, engineers and contractors, deal with conflicts, and negotiate a consensus. Agent based technology seems to offer a set of powerful concepts which can be of great help to model and build these new design systems. The DESSYS research will help us to get a better insight in the way design of foundations will change due to these developments in information-technology. Important goals are optimizing the design-process of foundations and integrating knowledge of the geotechnical engineer in the early design phase of the multidisciplinary design-process of buildings.

1 INTRODUCTION

This research started in 1996 when a project called 'Geotechnical knowledge, integrated in VR-DIS' (Van Veghel 1997) was initiated at the Eindhoven University of Technology in the Netherlands. During that project, a concept was developed concerning the design of foundations. The basic assumption was that because of the latest developments in Information Technology the design-process of foundations would change. The goal of the project: creating a concept that supports this basic assumption and develop a prototype to present it.

The concept was *founded* on the statements that:
- Design in the early design-phase was done by using geotechnical information from a (national) database
- Soil-investigation was based on the foundation pre-design.
- The designer was guided through the design process by means of a design supporting information system

The concept was supported by a prototype (Fig. 1) by means of which it was easy to explain the new concept. The concept was marked as promising both by the Eindhoven University of Technology as Fugro Ingenieursbureau B.V. in the Netherlands, which led to a joined PhD-research project in which the concept could be elaborated and developed further. The research is called 'DESSYS-knowledge modelling for a geotechnical design decision support system' (Van Veghel 1998). The research is primarily done at the Eindhoven University of Technology but the researcher is also working at Fugro Ingenieursbureau B.V. in order to get close to the domain. Fugro Ingenieursbureau B.V. is part of the worldwide operating Fugro N.V. holding company and provides geotechnical engineering and consulting services in the Netherlands. In the paper Fugro ingenieursbureau B.V. will be abbreviated by FUGRO.

In the concept, we can identify two major themes/components.
- *Workflow-management component*: DESSYS fits into the workflow-management system of FUGRO, which exists in a conceptual state of development.

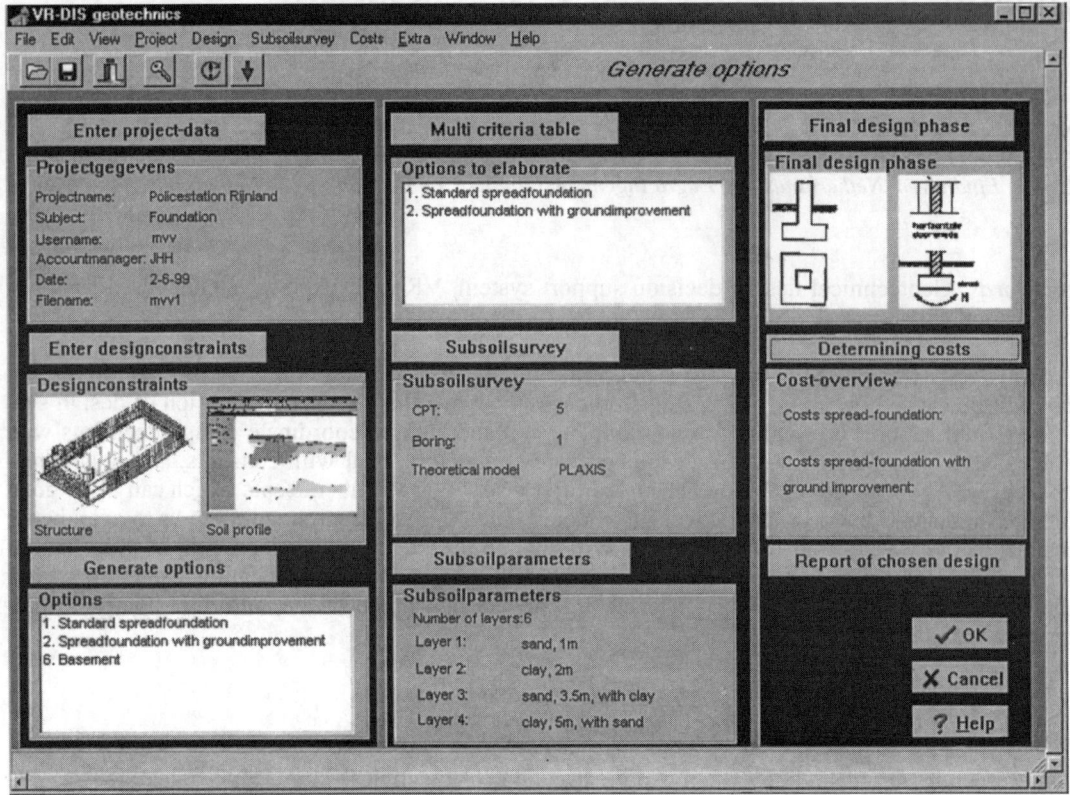

Figure 1. Screendump of prototype

– *Design decision support component*: DESSYS fits into the encapsulating VR-DIS (virtual reality design information system) research at the Eindhoven University of Technology. In VR-DIS, a multidisciplinary design system is developed using agent- and VR-technology.

2 DESSYS- KNOWLEDGE MODELLING FOR A GEOTECHNICAL DDSS

In Figure 2 is illustrated how the DESSYS-research fits into the VR-DIS research-program and the FUGRO workflow-management project.

2.1 *Workflow-management-component*

The course of the design-process as presented by the prototype was based on theoretical sources and has to be fine-tuned on real world daily practise. In 1998 FUGRO started a Business Process Redesign project in which one of the key issues was to analyse daily practise and think of a more efficient way to support it by means of IT-tools. At FUGRO the researcher is project-manager of this project. The knowledge derived from this project can be used in the research. In the left column of Figure 2 the workflow of projects at FUGRO can be identified. Projects start with a request from a client for a quotation and ends with archiving it. The DESSYS-research fits into the part where the calculating and designing of the foundation is being done.

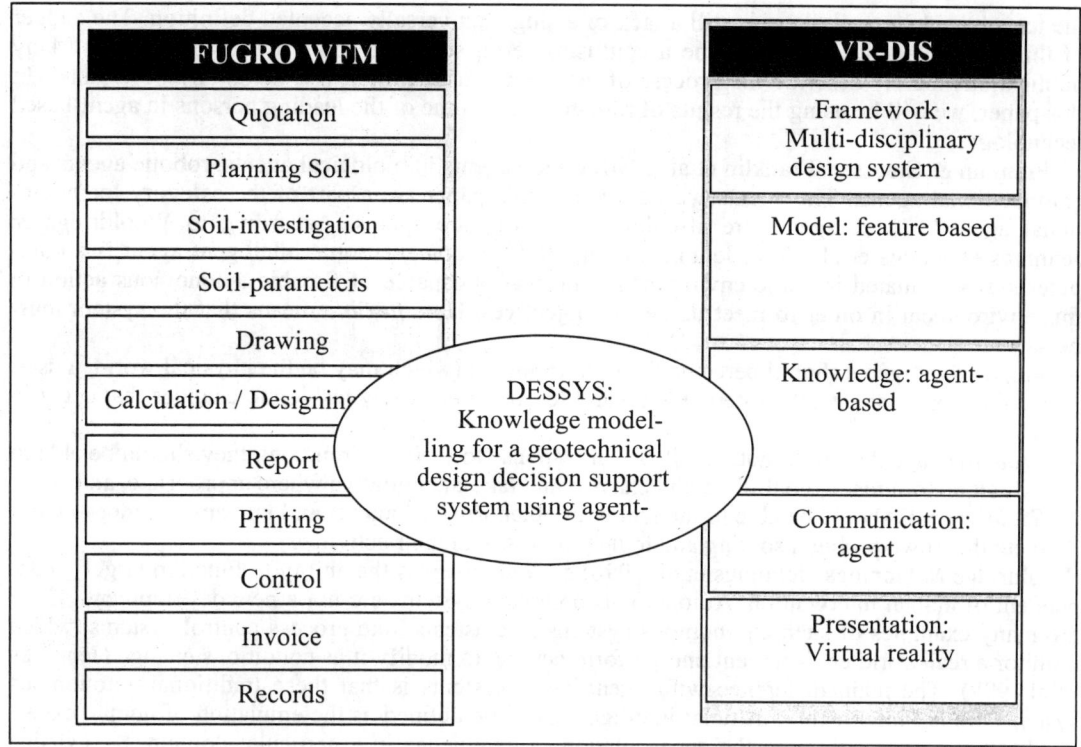

Figure 2. DESSYS within FUGRO WFM & VR-DIS.

2.2 *Design Decision Support-component*

The main goal of this part of the research-project is to integrate certain knowledge about designing foundations in a decision supporting design-system. Within the design-systems group of the Eindhoven University of Technology the project fits into the encapsulating VR-DIS research program, which is an abbreviation for Virtual Reality Design Information System. In this context the VR-DIS program and the DESSYS-research can be illustrated by Figure 5. VR-DIS can be identified as a multi-agent system in which different, so-called, agents represent the knowledge of the participants in the multidisciplinary design-process of a building.

By means of this multi-agent system different participants that are working on the design of a certain building can communicate and co-operate supported by their own agent. The scope of the DESSYS research is to develop the agent which represents/supports the geotechnical engineer. This way the knowledge of the geotechnical engineer is represented in the early stages of the design-process and not only after the actual design is finished. Probably the most interesting aspect of the concept is deriving foundation-solutions out of a number of design-criteria given by the physical situation and the architect.

3 AGENT-BASED DESIGN SYSTEM ARCHITECTURE

3.1 *Defining the term agent*

First, it has to be clear how we define the term 'agent' in this paper. The problem is that although the term is widely used, there is still a lack of a single universally accepted definition. The danger of this is that 'agent' might become a confusing term, subject to both abuse and misuse. Many publications merely describe the process of developing a definition and taxonomy of 'agents'. In this paper, we will be using the results of research from some of the leading persons in agent-based technology.

Franklin & Graesser (Franklin et al. 1996) divide agents in biological agents, robotic agents and computational agents. The agents we describe in this paper are a part of the category 'computational agents'. These agents are also known as software agents. According to Wooldridge & Jennings (Jennings et al. 1996, Jennings et al. 1998), a computational intelligent agent is a computer system situated in some environment, and that is capable of flexible autonomous action in this environment in order to meet its design objectives. Here *flexible* means that the system must be:
- *Responsive*: agents should perceive their environment (which may be the physical world, a user, a collection of agents, the Internet, etc.) and respond in a timely fashion to changes that occur in it,
- *Proactive:* agents should not simply act in response to their environment, they should be able to exhibit opportunistic, goal-directed behavior and take the initiative where appropriate, and
- *Social*: agents should be able to interact with other artificial agents and humans in order to complete their own problem solving and to help others with their activities.

Wooldridge & Jennings (Jennings et al. 1996) see *autonomy* as the ability to function largely independent of human intervention. Autonomous computer systems are not a new development. There are many examples of such autonomous systems in existence like process control systems, which monitor a real-world environment and perform actions to modify it as conditions change (Jennings et al.1998). The main difference with agent-based systems is that these traditional autonomous systems are not intelligent. Artificial intelligence can be defined as the emulation of human expertise by the computer through the encapsulation of knowledge in a particular domain. Knowledge can be predefined during the development of an agent but Russell argues that knowledge can also be attained during operation. According to Russell [Russell95] the term *autonomy* stands for the ability of an agent to base its behavior on both its own experience and the built-in knowledge used in constructing the agent for the particular environment in which it operates. Observations show that in general, intelligent agents have some initial knowledge as well as an ability to learn. This concept of *flexibility* and *autonomy* seems to be very promising because autonomy not only fits in with our intuition, but it is an example of sound-engineering practice. An agent that operates on the basis of predefined built-in assumptions will only operate successfully when those assumptions hold, and thus lack flexibility (Russell et al.1995). Hereafter, when the term agent is used in this paper, it should be understood that we are using it as an abbreviation for 'computational intelligent agent'.

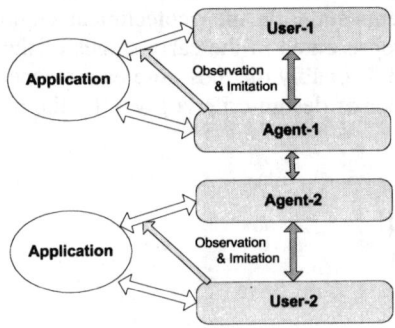

Figure 3. Multi-agent-system.

Some research focuses on developing sophisticated individual agents, mono-agents, with advanced capabilities, while other research is focused on multi-agent systems (Fig 3). A multi-agent system is designed and implemented as several interacting agents (Jennings et al. 1998).

The research on multi-agent systems is particularly focused on:
– Co-operation and collaboration;
– team- and coalition formation.
– information-sharing among the team;
– joint beliefs, goals and plans.

3.2 Multi-Agent system

The design-system we have in mind can be classified as a multi agent system. The research on multi-agent systems is particularly focused on co-operation and collaboration, team- and coalition formation, information sharing among the team, and joint beliefs, goals and plans. Within these fields, communication and Information Discovery are of significantly importance. Information Discovery (ID) is the synthesis of Information Retrieval and Information Filtering. The daily design practice requires a great deal of information to exchange between collaborators and the outside world. To support the designer to cope with this huge amount of information, agents can be used. In ID, there are three types of agents: user agents, broker agents, and resource agents (Wondergem et al. 1997).

User agents, also called user interface agents (Lashkari et al. 1998), satisfy the users' information needs. A user agent is aware of a user model, which describes the information needs of the user. This knowledge is used for retrieval and filtering. Each user agent learns by continuously 'looking over the shoulder' of the user as the user is performing actions. The user agent monitors the actions over a long period of time, finds recurrent patterns and offers to automate them. Broker agents form intermediaries between user agents and resource agents. They provide brokering services by matching user requests with the offered information by resource agents. Resource agents represent sources of information and are capable of indexing the available information. That is, they find and extract suitable information from CAD-models, documents, databases and external information resources like the Internet. In this phase of the research all these information sources are some how stored in what we call a virtual prototype. This virtual prototype is partly based on feature based modeling (Leeuwen et al. 1998) and acts like a blackboard. People only communicate with their personal user agent and user agents only communicate with resource agents via broker agents. Resource agents can communicate with each other. It is possible that they form combined packages of information. In addition, broker agents can exchange information about resources with each other (dotted arrow in Fig. 4). By learning, a user agent can accumulate so much knowledge, that it becomes to some extend an equivalent of the person it is attached to. If this occurs, we say that the user agent has become a knowledge-based agent. Whether this will happen, depends of the learning and memory capabilities of the agent.

4 THE DESSYS RESEARCH

4.1 Introduction to the DESSYS-research

Figure 5 illustrates the course of the DESSYS-research and the position of DESSYS in the VR_DIS research program. The DESSYS-research is part of the overall VR_DIS research-program of the Design-Systems group at the Eindhoven University of Technology. In the VR_DIS research program (Zutphen et al. 1996) an architectural design system will be built based on a virtual prototype of a building, with a feature based data-structure (Leeuwen et al. 1998) and using virtual reality techniques (Coomans et al. 1998).

The DESSYS-research covers knowledge modeling for a geotechnical design decision support

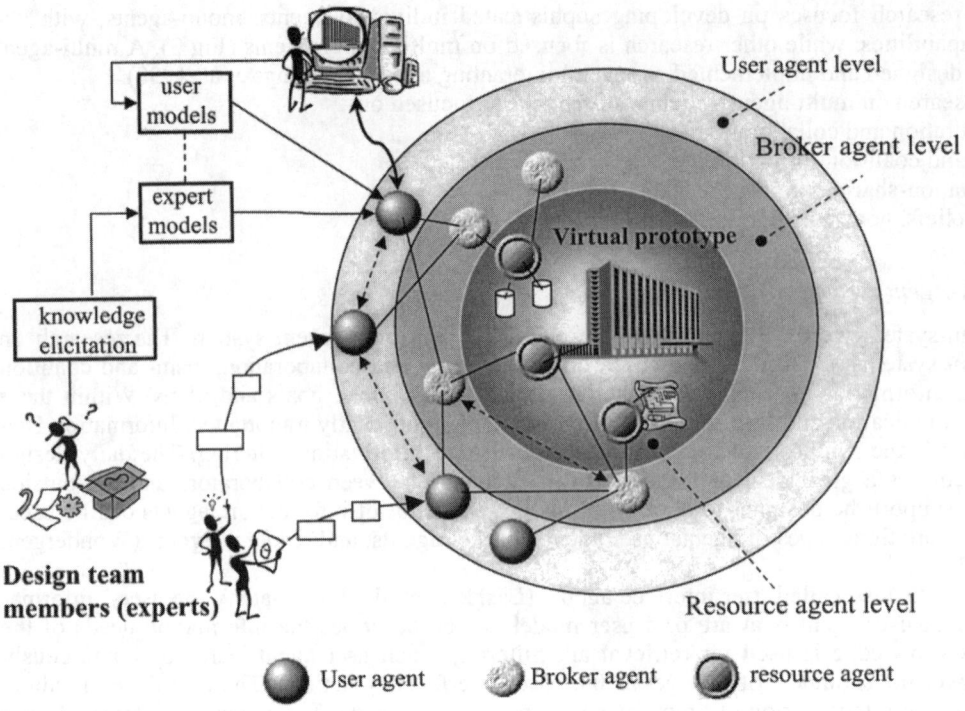

Figure 4. DESSYS within VR-DIS.

system. The course of the research is illustrated in Figure 5 and is divided in three phases. In the first phase of the project we will gain insight in the different techniques of transferring knowledge and how this is used in the process of making decisions in the structural-design. The field of study for this research will be the area of geotechnical engineering, foundation-design in particular. In the second phase we will gain insight in the formalization of expert-knowledge in such a way that it can be used as design-knowledge in the early phases of a multidisciplinary building-process. In the final phase the formalism will be validated by means of developing a knowledge-based agent which is an essential part of an agent-based design decision support system.

In Figure 5 two major directions regarding to agents can be distinguished. First, the course of building a mono-agent that represents one particular discipline/expert of the building-process. In the DESSYS-research (Veghel 1998) the discipline of geotechnical engineering is considered. This implies the design of the foundation of a building, so the agent represents the geotechnical expert. In Figure 5 this is illustrated by arrows from EXPERT A to AGENT A. Secondly, the interaction (multi-agent-system) and co-operation of different (singular) agents, representing different disciplines within the multidisciplinary process (like construction engineering, building-physics, structural design and architectural design) and of course the interaction with the user of the system (USER-AGENT A/AGENT A-AGENT B).

The design of a building is a matter of interaction between different parties, with their own concerns. We could think of this process as the interaction between different *collaborating agents*. Nwana (Nwana 1996) states that in order to have a coordinated set-up of collaborative agents, they may have to negotiate in order to reach mutually acceptable agreements. Why are collaborating agents a suitable metaphor for the design process? In other words: what is the motivation for having collaborative agent systems? Firstly, each single discipline is so complex that one centralized

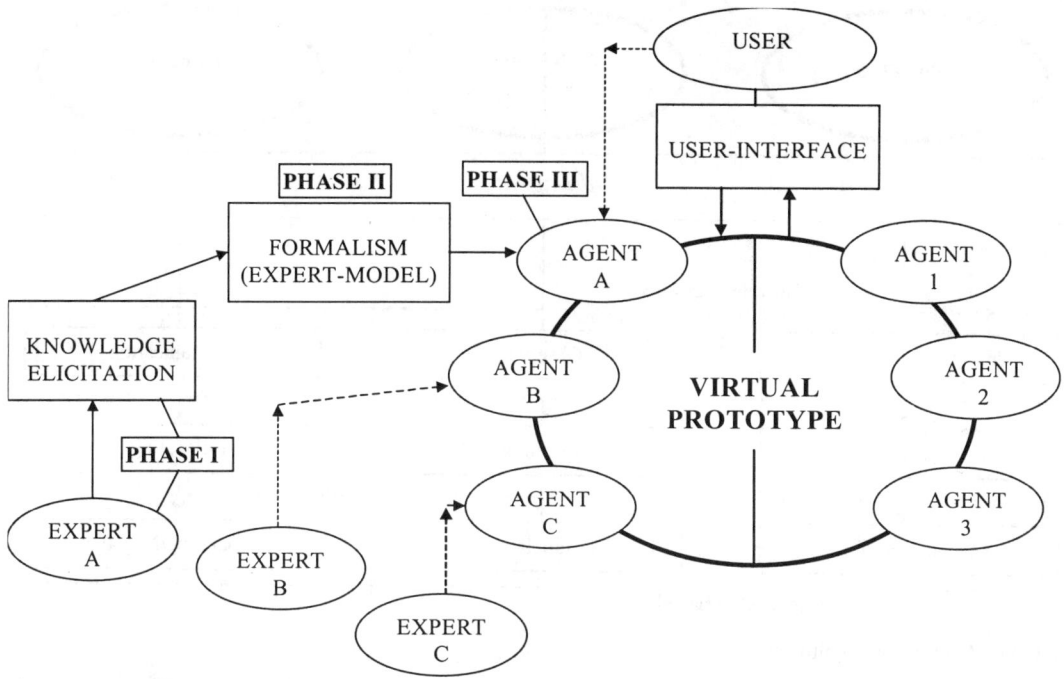

Figure 5. DESSYS & VR-DIS.

single agent for all disciplines is too hard to handle. Secondly, collaborative agent systems enhance modularity, which reduces complexity, speed, reliability, flexibility and the reusability.

So our agent A is a hybrid agent according to the typology of Nwana & Ndumu (Ndumu et al. 1997), consisting of two or more agent-philosophies. Our, hybrid, agent A is a mono-agent, which is part of a multi-agent system. So on the one hand its tasks are descended from the mono-agent – function while on the other hand its tasks are descended from its functioning in the multi-agent environment. We state, according to Nwana (Nwana 1996), that three primary attributes are important. A software-agent is an autonomous, goal directed process (Autonomy in Fig. 6). Further it is situated in, is aware of, and reacts to its environment (Adaptation in Fig. 6), and Co-operates with other software or human agents to accomplish its tasks (Co-operation in Fig. 6).

In Figure 6 the three basic agent-characteristics are translated in tasks that our agent should be able to execute. From top to bottom these tasks are positioned in an evolutionary order.

4.1.1 *Autonomy: providing information – assisting – making decisions*
The tasks derived from autonomy are mostly related to the function of mono-agent. The goal of the mono-agent is to become an autonomous assistant to the user, representing the knowledge of the expert, capable of making decisions. In the first instance the agent starts of with providing information to the user that is stored in a knowledge base; functioning as an expert-system. The agent gathers more information and knowledge from experience coming from prior cases and input from the expert or user (see 4.1.2 adaptation). Along the way providing information evolves to actively assisting the user in the process of design, based on stored information and knowledge. As the agent further evolves it will be able to autonomously make decisions on behalf of the user.

4.1.2 *Adaptation: observing – learning - manipulating*
Adaptation can be divided in adaptation to the user and the expert on the one hand and adaptation to other, collaborating, agents on the other hand. Adaptation to user and expert are related to the

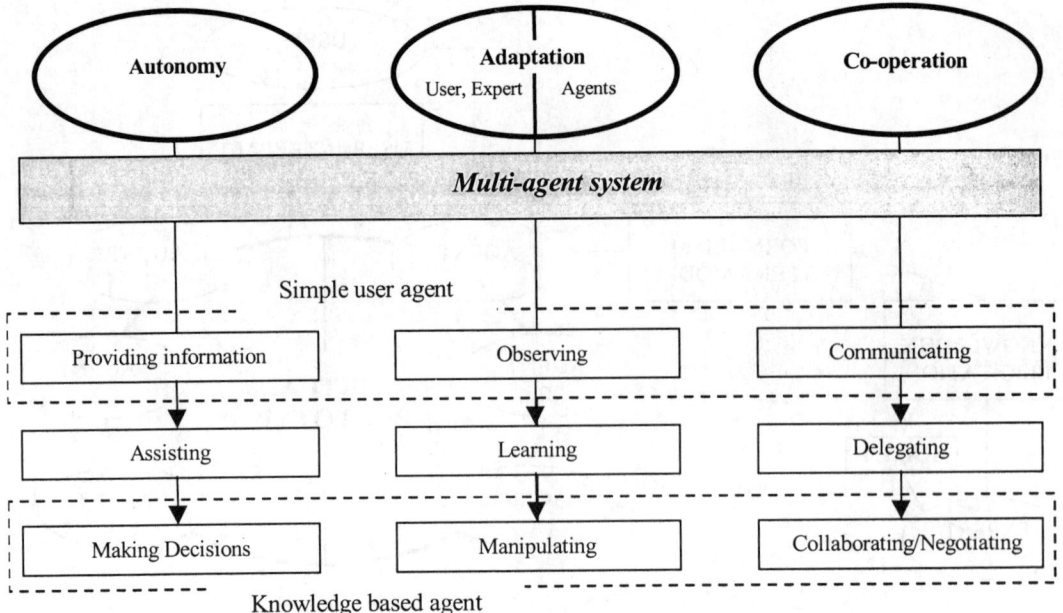

Figure 6: Agent tasks evolution.

mono-agent function, while adaptation to other agents relates to the multi-agent environment. In the first instance, the agent is merely observing its environment in order to act upon the stimuli that arise from it. The agent will be able to receive stimuli from other agents like i.e. a change in the structural design of the building and receive stimuli, input, from the user.
As our agent evolves, its task changes from merely observing the environment to learning from the environment. So the agent learns from the user by observing and analyzing the outcome of prior projects. It also learns how to react effectively to stimuli of collaborating agents from experience with them. Finally, the agent will be able to manipulate its environment supported by the things it has learned. For example, the agent will be able to send a message to the user that he needs more information or input. Another example is sending a message to collaborating agents that a certain kind of solution, regarding the design, would be an interesting option to consider.

4.1.3 *Co-operation: communicating – delegating – collaborating/negotiating*
The tasks derived from co-operation are mostly related to the function of the agent in a multi-agent environment. The intention of the 'multi-agent' is to become a 'respected' partner in the team of collaborating agents, perceiving a collective goal. Communication amongst each other is probably the most important skill for agents. *Communication in both ways is thé distinctive aspect that makes the difference between an agent and an ordinary program.* Therefore, from the beginning communication between different agents is necessary. As the agent evolves it can delegate tasks, to other agents or possibly to the expert. Eventually the agent will be able to, not only communicate and delegate but also, collaborate and negotiate with other agents and the user or expert. I.e. if agent A favours an option that conflicts with the preferences of agent B, both the agents should be able to collaborate on the level of negotiation. This imposes restraints on the language both agents communicate with.

4.2 *Creating an 'Knowledge based agent'*
The agent we desire acts as an 'intelligent assistant to the human-expert' for example an engineer

or an architect). It assists the human-expert during the design and engineering of the foundation of a building or structure, based on expert-knowledge. One could call it an 'knowledge-based agent'. Making knowledge of experts attainable is a function of a traditional expert system. The difference between and knowledge based agent and an expert system is that you could consider the knowledge-based agent to be a pro-active expert system interacting with the user, learning from the user. The DESSYS-research will investigate the possibility of creating such a knowledge-based agent.

4.2.1 Characteristics of the 'knowledge based agent'

The DESSYS-research will investigate the use of an agent (agent A in Fig. 5) which represents the geotechnical expert during the process of designing the virtual prototype of a building. The tasks that can be executed by an agent in this process are described below. Firstly, we look at the characteristics. As a matter of course, this agent will incorporate the three basic agent characteristics: Autonomy, Adaptation & Co-operation (Nwana 1996). In this context these characteristics imply:

- *Autonomy*: the agent will act autonomously to pursue its goals, which in this research will be the design and calculation of a foundation of a building. The agent will act both proactive and reactive. It will be reactive when it reacts directly on input from the user. It will be proactive when it takes the initiative whenever the situation asks for it. For example: the agent will inform the user of an awkward choice during the design and will provide a better alternative.
- *Adaptation, learning*: the agent adapts to its environment and users and learns from experience. The agent's behavior can be based on both its own experience and the built-in knowledge. Therefore, it is necessary to provide the agent with some initial knowledge as well as an ability to learn. For example: the agent should be able to accelerate the process of finding a basic-solution of the design problem, if ground-conditions and other preconditions are much like a situation, which earlier has been coped with.
- *Co-operation*: the agent (Agent A in Fig. 5) will co-operate and collaborate with other agents (Agent B,C, et al. in Fig. 5) to achieve common goals. In this research, agents should collaborate on the level of the virtual prototype. The common goal of the collaborating agents is designing a virtual prototype of a building, which meets the requirements in the inception-phase with an optimum ratio of cost to quality. Co-operation is necessary in this context because it enables:
  - Meeting global constraints that cannot be met by any one agent acting in isolation;
  - Maximizing utility through the sharing of distributed expertise, resources or information.

4.2.2 Agent-based design of a foundation

This section will describe the role agents can play in the process of designing and engineering a foundation for a building or structure. This description will be supported by illustrations, which are derived from an interface-prototype from Van Veghel (Veghel 1997).

In the design of a foundation, according to Van Veghel, we distinguish three stages:
- Generating a number of variants on the level of pre-design of the foundation;
- Choosing which variants to elaborate;
- Elaborate at least one variant.

The general concept of the theory is to develop the pre-design of the foundation on the bases of already available geotechnical data of the sub-soil. Subsequently the subsoil-survey can be based on the pre-design. This way it is certain that the exploration-points are taken on the right spot.

I. Generating a number of variants on the level of pre-design of the foundation

In the first stage, the agent needs to generate a number of variants that are suitable for the given situation. The agent can only do this if it has sufficient information about the structural design of the building, the soil-conditions, the loads, the building-site and the environment.
- *Structural design of the building*. Here we encounter the first possible collaboration with another agent, the agent that represents the structural design. The data that is needed in this stage of the design can be obtained via input from the user or, better, via interaction from one agent, the structural-agent, to the other, the foundation-agent.

122 M. van Veghel

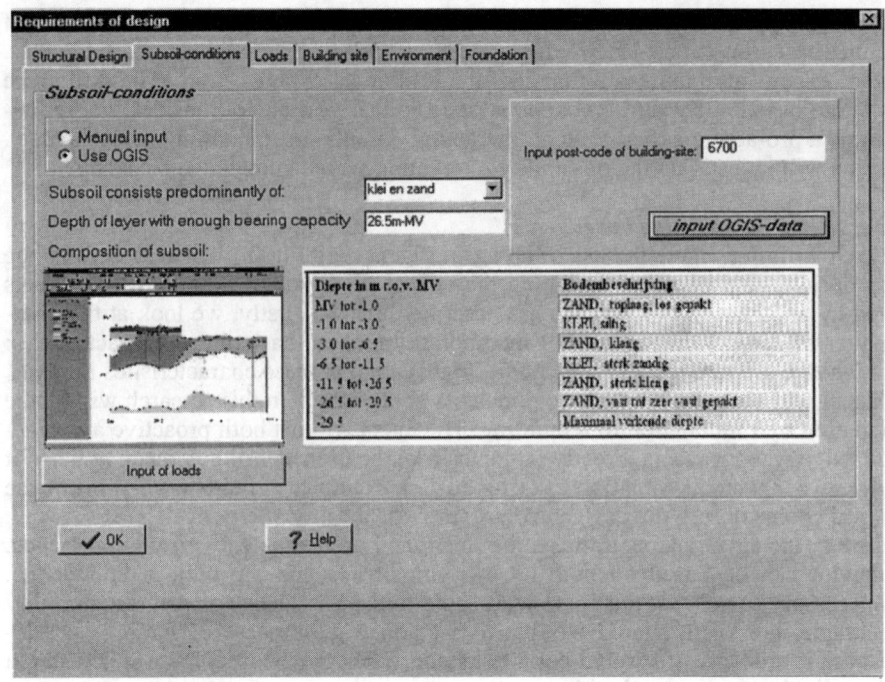

Figure 7: Design criteria

- *Subsoil-conditions*. It is obvious that the soil-conditions of the sub-soil of the building are very important. These conditions imply to great extend the possibilities of certain foundation-solutions. In this stage of the design, the geotechnical data is retrieved from a database, which can be internal or external. The agent knows the location of the building-site from the user or from collaborating agents. The agent will start looking for the proper information, first in the internal database and subsequently in the external databases via the Internet. An agent with this behavior is called a mobile-agent.
- *Loads*. The dead weight of the building combined with the different types of loads is decisive for the (pre) design of the foundation. In the stage of pre-design, the load can be roughly calculated, depending on the function of the building
- *Building site*. Here we encounter another possible collaboration with another agent, the agent that represents the construction-specialist. The dimensions, the soil-conditions and the ease to reach the building-site are effecting the process of generating variants. This information can be provided by the user or, better, through interacting of our agent with the construction-agent.
- *Environment*. The environment of the building site plays an important role in the design of the foundation. The admissibility of vibrations, noise and changes of the groundwater level are restricting conditions.

At this point, the agent is aware of all the information it needs in order to generate a number of variants. The ideal knowledge based agent, regarding to knowledge:
- Has access to background knowledge, which is available and general
- Can be controlled by the user, especially when the agent is new or drastic changes occur in the user's behaviour
- Learns to adapt & suggest changes

In the first instance the agent will generate the variants purely based on pre-programmed knowledge. During the process of operation, the agent will learn from the user. By looking at the outcome of the process of design, the agent can adjust the way it generates its variants. So when an apparent similar situation occurs, the agent will relate it to preceding cases. In this way it is adapting to the user and combining this experience with its pre-programmed knowledge.

II. Choosing which variants to elaborate

There is a significant difference between the nature of the tasks in stage I and in this stage. The tasks in stage I lend itself to be done by an agent. Figure 8 illustrates a possible result of stage I. The agent has generated six variants, which in the given situation could be applied. The agent bases this on the given pre-conditions. Next, the user chooses which variants to elaborate on the level of preliminary-design. Now the agent calculates the chosen variants and composes a multi-criteria table. This table indicates a score on a certain scale on various criteria for each chosen variant.

For example the score of a variant on the use of space on the site or on environmental criteria like noise, vibrations and disturbance of the groundwater level. Next, the user can decide, on the basis of the table and his priorities, which variant or variants to elaborate. In this stage, the agent is typically acting as an assistant to the user.

III. Elaborate at least one variant: final design stage

The result of stage II is a preliminary design of one or more foundation-variants. Subsequently the soil-investigation is based on the pre-design. This way it is certain that the test-locations are taken on the right spot. After soil-investigation, the final design stage can be entered. In this stage the interaction between our agent and the agent or expert representing the structural design is crucial. The design of the foundation is strongly related to the structural design of the building. This interaction between these two disciplines is often neglected in current engineering practise, because of its complexity. Collaborating agents could be a solution to this problem.

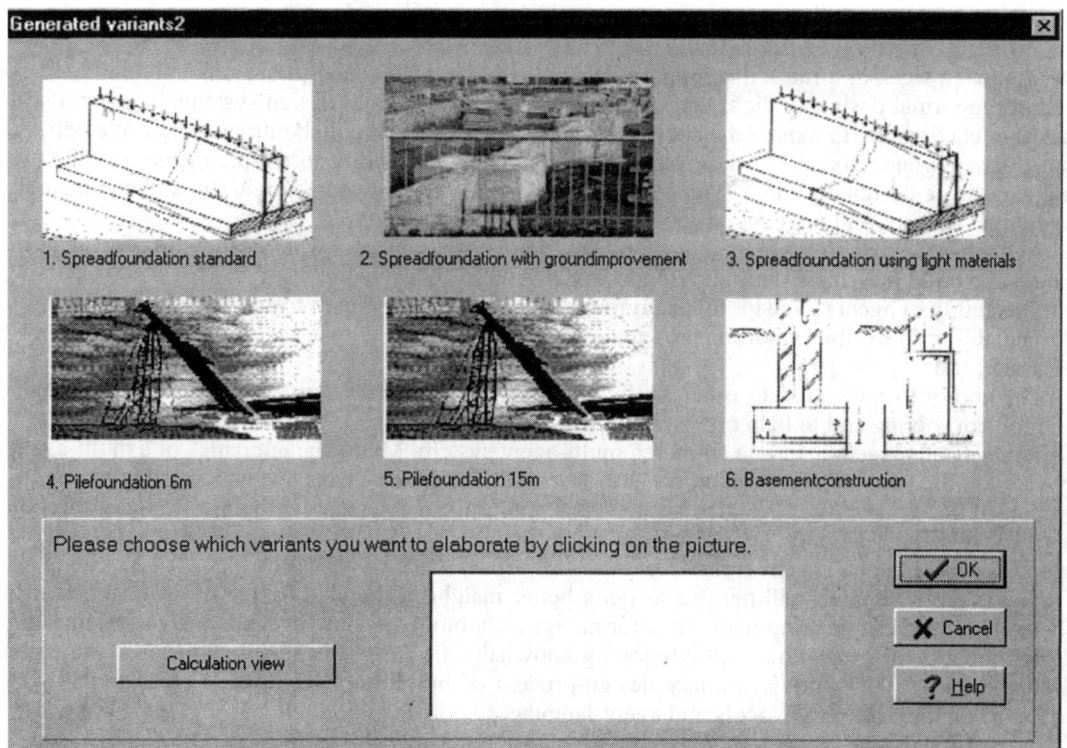

Figure 8. Generated variants.

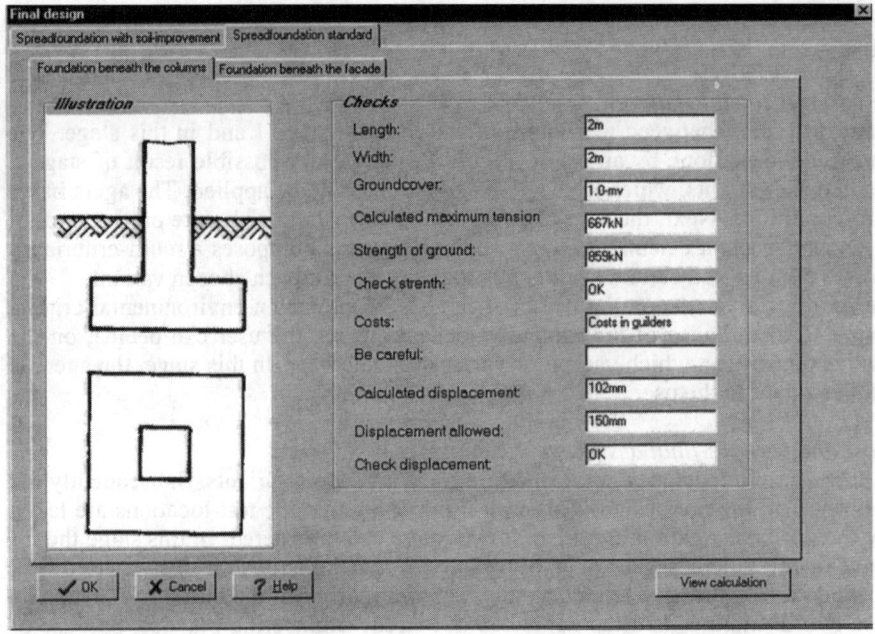

Figure 9 : Final design

5 DISCUSSION

Design is the work of a multi disciplinary design team that in the near future probably will be carried out in virtual design studios. We argued that CAD-systems and design systems in general still lack the functionality to handle the complex, dynamic character of collaborative design. We believe that the next generation of design system should support the ability to communicate, collaborate and coordinate design decisions with clients, users, legislators, advisors, engineers and contractors, deal with conflicts, and negotiate a consensus.

Agent based technology seems to offer a set of powerful concepts which can be of great help to model and build these new design systems. Some of these concepts are
- The ability of agents to adapt to its environment and users, and learn from experience,
- The ability to exhibit opportunistic, goal-directed behavior and take the initiative where appropriate, and
- The ability to interact with other artificial agents and humans in order to complete their own problem solving and to help others with their activities.

Different agents together form a so-called multi-agent system. Main characteristics of a multi-agent system are the support of co-operation and collaboration, exactly what we need for the next generation of design systems. We also showed that multi-agent systems can help a designer to cope with today's huge amount of information by utilizing user agents, broker agents and resource agents for Information Discovery.

The DESSYS research will help us to get a better insight in the way design of foundations will change due to these developments in information-technology. Important goals are optimizing the design-process of foundations and integrating knowledge of the geotechnical engineer in the early design phase of the multidisciplinary design-process of buildings. Our next research topics will concern communication protocols and agent-languages.

6 ACKNOWLEDGEMENTS

This research is performed within the VR-DIS research program, and builds on the Ph.D. research by Mr. M. van Veghel under supervision of Prof. Dr. H.J.P. Timmermans and Associate Prof. R.H.M. van Zutphen from the Eindhoven University of Technology, and Prof. A.F. van Tol from the Delft University of Technology. Thanks are due to Mr. L. de Quelerij from Fugro, a multinational group of consulting engineers for their co-operation and funding of this research. Information of the research and the researcher can be found at http://www.ds.arch.tue.nl .

REFERENCES

Coomans, M.K.D. & Timmermans, H.J.P. 1998. 'A VR-User interface for design by features', Eindhoven University of Technology, Proceedings of the 4th Conference on Design and Decision Support Systems in Architecture and Urban Planning. Maastricht, the Netherlands, July 26-29, 1998 (forthcoming).
Franklin, Stan & Graesser, Art 1996. 'Is it an agent, or just a program? A taxonomy for autonomous agents', Proceedings of the 3rd international workshop on agent theories, architectures, and languages. Springer-Verlag.
Harrison, C.G., Chess, D.M. & Kershenbaum, A. 1995. 'Mobile agents: Are they a good idea?', IBM T.J. Watson Research center.
Jennings, N.R. & Wooldridge, M. 1996. 'Software agents', IEE Review: 17-20.
Jennings, N.R. & Wooldridge, M. 1998. 'Agent-technology, foundations, applications and markets', Springer Computer Science.
Lanier, J.. 'Agents of alienation', http://www.voyagerco.com/consider/agents/jaron.html.
Lashkari, Y., Metral, M & Maes, P. 1998. 'Collaborative interface agents' in Huhns, M.N. & Singh, M.P. (eds) 'Readings in agents'.
Lawson, B. 1997. 'How designers think, the design process demystified', Oxford Architectural Press, 3rd edition.
Leeuwen, J. & Achten, H. 1998. 'A feature based description technique for design processes: A case study', Eindhoven University of Technology, Proceedings of the 4th Conference on Design and Decision Support Systems in Architecture and Urban Planning. Maastricht, the Netherlands, July 26-29, 1998 (forthcoming).
Maes, Patti. 'Software-agents'-slide show: http://pattie.www.media.mit.edu/people/pattie/CHI97/sld001.htm.
Mitchell, W.J.. 'Creative Design in the Computer Era-The design Studio of the Future', MIT School of Architecture and Planning, http://sap.mit.edu/dsof/research/creative_design.html.
Ndumu, D.T. & Nwana, H.S. 1997. 'Research and development challenges for agent-based systems', IEE proceedings.
Nwana, Hyancinth S. 1996. 'Software agents: an overview', Knowledge engineering review, October: 205-244.
Petrie, C.J. 1996. 'Agent-based engineering, the web, and intelligence', Stanford center for design research, IEEE Expert.
Russel, Stuart J. & NORVIG. P. 1995. 'Artificial Intelligence: A Modern Approach', Englewood Cliffs, NJ: Prentice Hall.
Van Veghel, M.M.P.H.L. 1998. 'DESSYS, knowledge modelling for a geotechnical design decision support system' research-proposal, Eindhoven University of Technology.
Van Veghel, M.M.P.H.L. 1997. 'Integratie funderingsontwerp in VR-DIS' (in Dutch) , Eindhoven University of Technology.
Wooldridge, M. & Jennings, N.R. 1995. 'Intelligent agents: Theory and Practice', Knowledge engineering review 10(2): 115-152.
Wondergem, B.C.M., van Bommel, P. & Huibers, T.W.C.. 'Agents in cyberspace : towards a framework for Multi-Agent Systems in Information', Katholieke Universiteit Nijmegen, CSI-R Computing Science Institute report 9715.
Zutphen, R.H.M. & Mantelers, J.M.M. 1996. Computational design: simulation in Virtual Environments. *Proceedings of the 3rd Conference on Design and Decision Support Systems in Architecture and Urban Planning*. Spa, Belgium, August 18-21, 1996.

Miscellaneous

Aspects for dynamic compaction of saturated sand

P. VUOLA
VR-Track Ltd, Helsinki, Finland

J. HARTIKAINEN
Tampere University of Technology, Tampere, Finland

ABSTRACT: Dynamic compaction of saturated sand by heavy tamping was researched. One interest in the research was, whether stratification, especially rock in a finite depth, affects compaction depth. Another interest was, if different ground vibration wave types can be distinguished from each other and analysed analytically. As a result a modified compaction depth equation is presented. For environmental impact assessment statistical test result analysis for predicting vibrations on the ground surface in a homogeneous half space is presented.

1 INTRODUCTION

Heavy tamping is a method, which is used to compact soil deep below the ground surface by dropping a heavy block from a height. The term arises from the point of work performance. Heavy tamping has proved to be technically available and economical, and it has its own share in the market.

Dynamic compaction is a term, which describes soil behaviour during compaction seen from inside the soil. An essential behaviour of soil during dynamic compaction is that compaction process in the ground is immediate.

Environmental impact as vibrations is induced into the ground and thus to the environment by heavy tamping. Vibrations are waves in the ground. In many cases the environment reduces the utilization of the method.

The dominant method to predict compaction depth is the Ménard and Broise equation based on energy. The equation is semi empirical, and it has been derived for a half space, i.e. for such a case, where the ground is homogeneous and soil deposit very thick. Anyway, a real ground is seldom a half space. For example in Finland rock is quite close to the ground surface, especially in many areas, where heavy tamping is a potential ground improvement method.

Soil vibration prediction methods are mostly empirical. Vibrations are often presented as if there were only one wave type, though three different wave types are known from wave theory.

Two major questions were asked:
a. Does rock in a finite depth affect compaction depth, and
b. Can wave types be distinguished from each other and analysed according to wave theories?

2 TEST PROGRAMME

2.1 Test sites

Test sites were situated in Satakunta, a province in western Finland.

Sites A1 and A2 were situated in an industrial area. Compacted crushed fill on the ground surface is called cushion in this text. Approximate cushion thickness was on site A1 0.2 m and on site A2 0.5 m. Soils on the sites are from a sedimentary origin: alluvial top deposits and glacial outwash deposits deep in the ground. Below the top loose sand layer soil is mostly silt and sand. Soil conditions are close to a half space. Ground water table depth was approximately 1.0-1.5 m, when heavy tamping was started.

Site B was situated in a harbour area. Approximate cushion thickness was 0.5 m. The loose sand layer is a hydraulic fill above bedrock. Ground water table depth was approximately 2.0-2.5 m, when heavy tamping was started.

Grain distribution curves of the saturated sands are presented in Figure 1.

2.2 *Compaction procedure*

Compaction device is presented in Figure 2. Block mass was m = 10.5 tons and drop height H = 12.5 metres. The drop was free. Drop grid was 2 m * 2 m in a square pattern. Two drop rounds were used. One round consisted of two drops on each grid point.

On site A2 altogether three compaction rounds were used. Compaction during the second round was essential, but any further compaction during the third round could not be distinguished from Weigth Sounding Test diagrams. So, two compaction rounds seem to yield to maximum compaction with this devise.

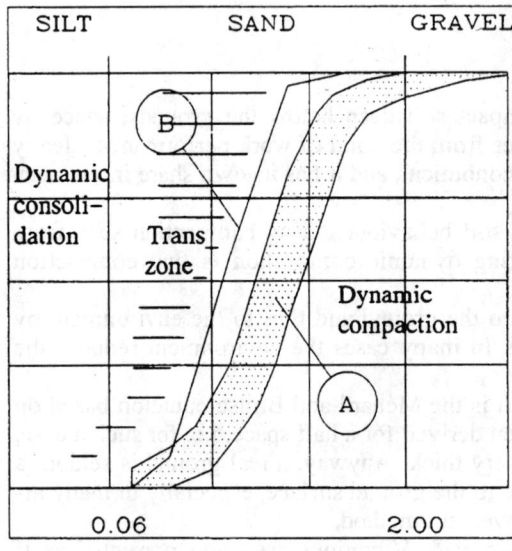

Figure 1. Grain distribution curves of the sands from sites A and B.

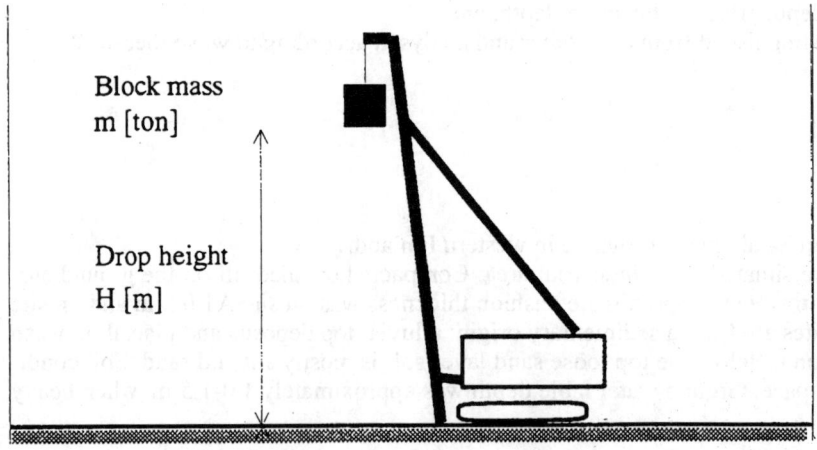

Figure 2. Principle of compaction device.

3 COMPACTION ANALYSIS

3.1 Test results

Block behaviour test results are presented in Figure 3. The dashed line is from site A1, the thin continuous line from site A2 and the thick line from site B. Block velocity was approximately 15.7 m/s in the beginning of impact at time 0 ms.

The term initial impulse is used in this research. It is defined to be the impulse, during which the block retardation has its maximum value and is approximately constant. Initial impulse duration varied between 5-10 ms. On sites A2 and B the initial impulse was approximately 40 % of the total impulse and on site A1 25-30 %.

Soil behaviour test results are presented in Figure 4. The penetration test method was Weight Sounding Test. A rough estimate of corresponding CPT values is presented as well.

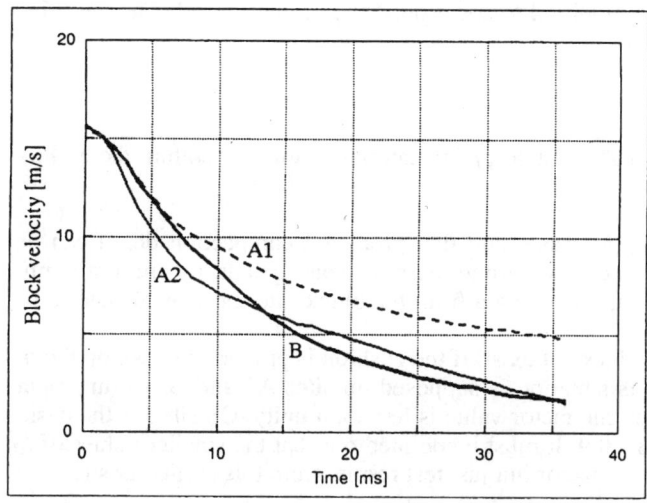

Figure 3. Representative block velocity curves during impact.

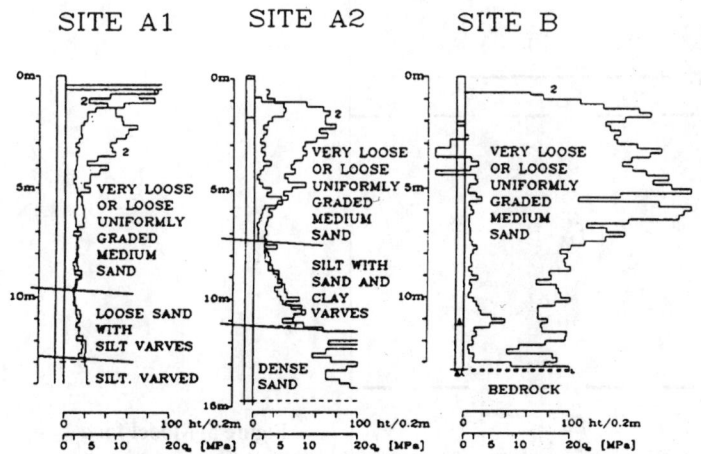

Figure 4. Dynamic compaction test results, two compaction rounds.

The cushion seems to affect maximum compaction depth to some extent according to sites A1 and A2. A finite rock depth seems to affect compaction depth remarkably on site B.

The traditional Ménard and Broise Equation (1) used in compaction depth calculations is

$$D = k * \sqrt{(m*H)} \qquad (1)$$

where D = depth [m], k = factor, m = block mass [ton], and H = drop height [m].

Test results from sites A1, A2 and B are approximately k = 0.6, 0.7 and 1.1, respectively. Literature values for sand are k = 0.5 – 0.8.

3.2 Modified depth Equation

The material, sand, was almost similar on all sites, but the traditional factor k varies remarkably according to the test results. The differences between the sites were the cushion and the bottom. A model for stratified, or layered, ground instead of a half space is presented in Figure 5.

The traditional energy equation can be modified to correspond the ground model. The modified maximum compaction depth Equation (2) is

$$D_{MAX} = f_C * f_M * f_B * \sqrt{(m*H)} \qquad (2)$$

where D_{MAX} = maximum depth [m], f_C = cushion factor, f_M = material factor, f_B = bottom factor, m = block mass [ton], and H = drop height [m].

Cushion factor. The cushion is necessary to make sure that predicted compaction depth can be achieved. The geotechnical purposes of the cushion are to increase bearing capacity and initial impulse (impulse during maximum retardation, transmitted from the block into the ground) and to be a drainage layer.

An analytical expression for the factor does not exist. If the cushion is proper, the cushion factor value can be determined to f_C = 1.0. This situation is supposed on sites A2 and B. An unproper cushion reduces maximum depth, and then the factor value is less than unity. On site A1 the cushion factor value is approximately f_C = 0.8 - 0.9. It must be pointed out that the smaller values of f_C from site A1 are not minimum values for the factor but just test results from this particular site.

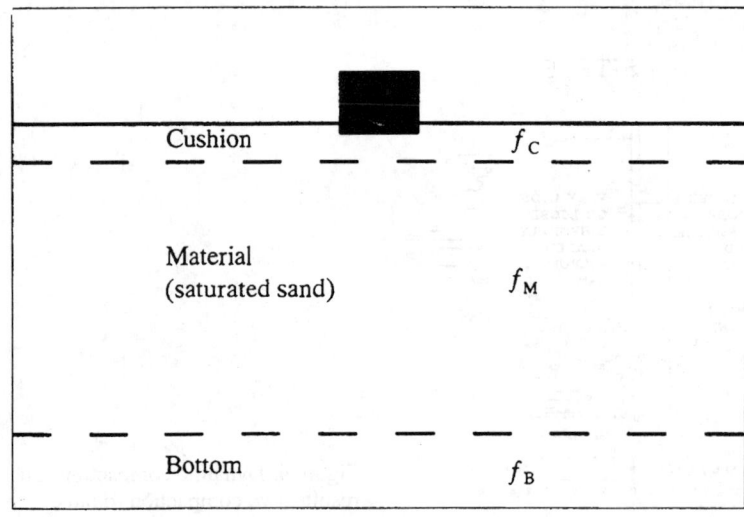

Figure 5. Model for a stratified ground.

Material factor. The numerical value for saturated sand is $f_M = 0.65 - 0.75$. These values have been gathered from several literature sources.

Bottom factor. Rock in a finite depth increases maximum compaction depth. This phenomenon is contributed by impact wave reflection from the bottom. Wave interference increases dynamic stresses and strains in soil and increases compaction depth.

Waves were analysed according to the theories by Zoeppritz, presented by Richart (1970). According to this theory almost 100 % of the impact wave is reflected from the rock surface. Besides reflection and wave interference, soil liquefaction potential and geometrical damping of the wave were considered.

According to theoretical calculations a bottom factor value approximately $f_B = 1.5 - 2.0$ was derived. From the test results a value $f_B > 1.5$ can be estimated. Apparently the rock surface was above maximum compaction depth and so the maximum value of f_B could not be determined.

Maximum depth. The Equation (2) and bottom factor values presented above apply, when the rock surface is so near that compaction takes place as far as the rock surface.

The factor values presented in this text are not any exact values suggested for engineering. Further field tests and analyses are necessary so that standard values could be recommended. Anyway, the tests and analysed results show that ground stratification affects compaction depth, and the depth equation modification is necessary.

4 ENVIRONMENTAL IMPACT

4.1 *Wave types, propagation and damping*

The major environmental impact induced by heavy tamping is vibrations. There is two kinds of body waves: P-wave or primary wave and S-wave or shear wave. The surface wave is called R-wave or Rayleigh wave. P-wave is the fastest and R-wave the slowest wave type. Wave types are presented in Figure 6.

Body wave propagation is hemispherical and R-wave propagation cylindrical. Geometrical damping of body wave particle displacement amplitude near the ground surface is proportional to $1/d^2$, where d is the hemisphere radius or distance from the vibration source. Geometrical

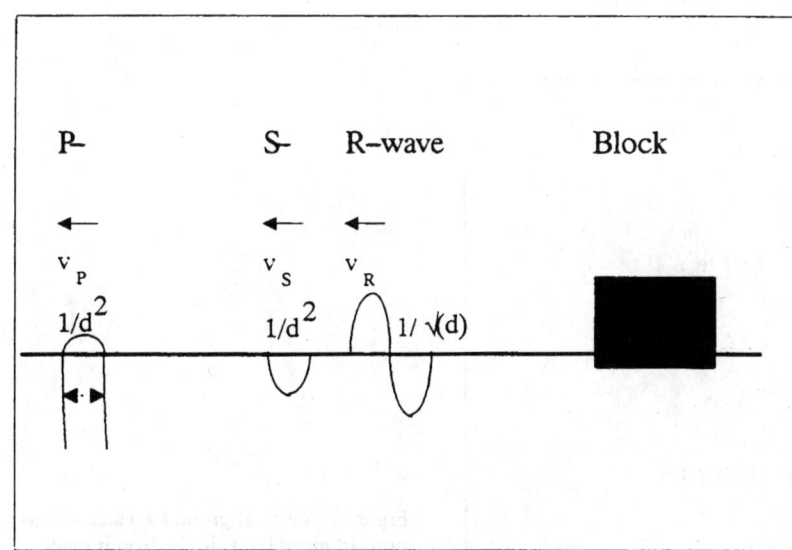

Figure 6. Wave types and geometrical damping.

damping of R-wave displacement amplitude is proportional to square root of (1/d), where d is the cylinder radius or distance from the vibration source. Principles of geometrical damping are presented in Figure 6.

The principles presented above are for a homogeneous isotropic half space. In a real ground reflection and refraction take place due to soil inhomogeneity, soil layer interface inclination and finite rock depth.

Besides geometrical damping there is material damping. It is energy transformation from kinetic to heat and plastic deformations.

4.2 Test result analysis

The test area was site A1. A representative vibration diagram is presented in Figure 7.

All three wave types can be distinguished in the diagram. Frequency of the R-waves varied from 5 to 7 Hz. Frequency of the S-waves varied from 5 to 7 Hz that of the P-waves from 8 to 13 Hz.

Vibration test results as maximum vertical particle displacement velocity are presented in Figure 8. Units used in the figure are: v = maximum particle displacement velocity [mm/s], d = distance [m], m = block mass [ton], and H = drop height [m].

In Figure 8 regression lines and statistical variation lines, $\pm 1*$ and $\pm 2*$ standard deviation, are presented. Maximum vibrations were induced by body waves or interference of body and Rayleigh waves in the steeper line area, which represents close ranges from the vibration source. Maximum vibrations were induced by R-waves in the more gentle line area, which represents great distances from the vibration source. This phenomenon was revealed from ground surface vibration diagrams as well as from statistical analysis presented in Figure 8. The transition point from body wave area to R-wave area was approximately d = 14 m in this test.

5 CONCLUSIONS

Soil stratification or layering affects compaction depth. Bedrock in a finite depth increases compaction depth at least by a half. The cushion, compacted crushed top fill, affects compaction depth to some extent. As a result a modified depth equation is presented.

Different types of environmental waves can be distinguished from each other in such a ground, which is close to a half space. Body waves dominate in close ranges and surface waves in greater distances. As a result statistical analysis is presented.

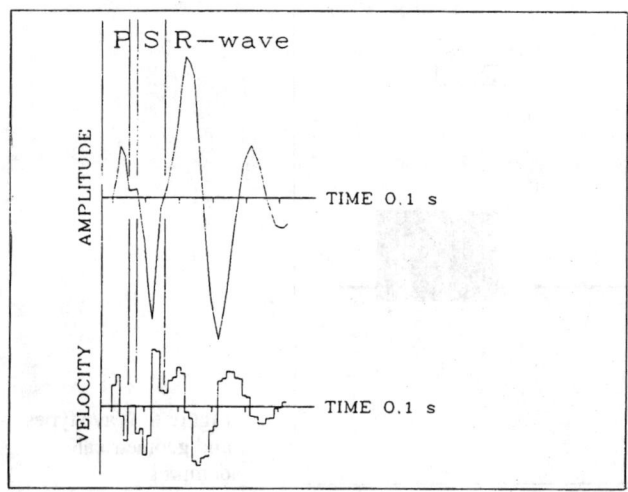

Figure 7. Vertical ground surface vibrations induced by a single drop impact.

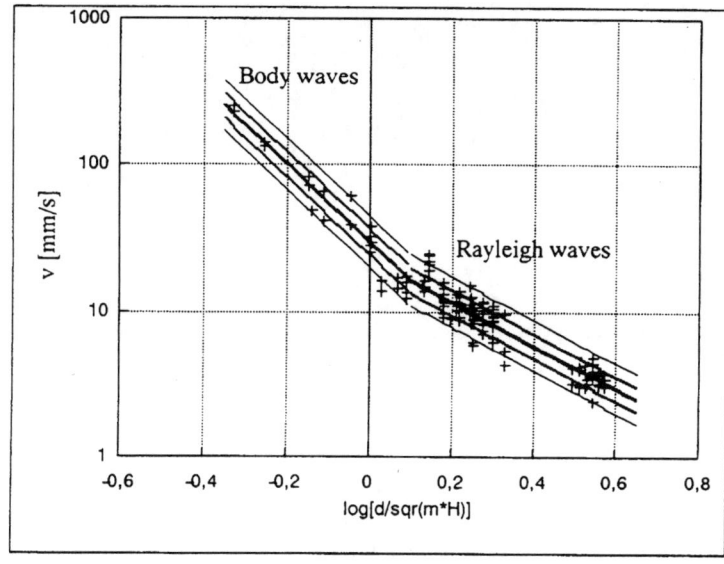

Figure 8. Statistical analysis of vertical vibrations on the ground surface for a thick soil deposit.

REFERENCES

Hansbo, S. 1990. *Jordförstärkning*. Svenskt Tryck Stockholm.
Hartikainen, J. 1986. Maapohjan vahvistaminen. *RIL 166 Pohjarakenteet*.
van Impe, W. & Madhav, M. 1995. *Deep dynamic Densification of granular Deposits*. Soil Mechanics Laboratory, Ghent University.
Mayne, P. et.al. 1984. Ground response to dynamic compaction. *Journal of Geotechnical Engineering*, ASCE 110(6).
Ménard, L. and Broise, Y. 1975. Theoretical and practical Aspects of Dynamic Consolidation. *Geotechnique*, 25(1).
Richart, F. Jr. et.al. 1970. *Vibrations of Soils and Foundations*. Prentice-Hall.
Sêco e Pinto, P. (ed.) 1993. *Soil Dynamics and Geotechnical Earthquake Engineering*. Van Impe et. al: Dynamic soil improvement methods. A.A.Balkema.
Smoltczyk, U. 1983. General report: Deep Compaction. *The 8th European Conference on Soil Mechanics and Foundation Engineering*. Helsinki.
Vuola, P., Hartikainen, J. & Hakulinen, M. 1995. Aspects for dynamic compaction of saturated sand. Workshop on Ground Improvement. *The 11th European Conference on Soil Mechanics and Foundation Engineering*. Kopenhagen. Unpublished.
Vuola, P. 1996. *Dynamic compaction of saturated sand*. Licentiate thesis. Tampere University of Technology, Geotechnical laboratory Publication 37.
Vuola, P. (Unpublished). *Bottom factor in maximum compaction depth equation for dynamic compaction of saturated sand*.

List of delegates of XII ECSMGE 1999, Amsterdam

Algeria
Bahar, Ramdane; Universite de Tizi-Ouzou, Tizi-Ouzou

Australia
Ervin, Max; Golder Associates, Hawthorn West
Hargreaves, Bruce; Efa Geotechnical, Eight Mile Plains
Indraratna, Buddhima; University of Wollongong, Wollongong
Nash, Clarry; Soil Testing & Design Services, Queensland
Raisbeck, Don; Kinhill, Melbourne, Vig
Randolph, Mark; The University of Western Australia, Western Australia

Austria
Adam, Dietmar; TU-Vienna, Vienna
Brandl, Heinz; Techn. Universitat Wien, Wien
Fross, Manfred; Vienna University of Techn., Wien
Fuchsberger, Martin; Technical University Graz, Graz
Goriupp, Harald; Amtd. Steierm. Landesregierung, Graz
Hazivar, Wolfgang; Wien
Kaltenegger, Stefan; Austrian Federal Railways, Wien
Kammerer, Georg; Technical University Graz, Graz
Leitner, Knut; Vienna
Lugmayr, Rainerd; Polyfelt, Linz
Martak, Lothar; Goverm. of the City of Vienna, Vienna
Schweiger, Helmut F.; Technical University Graz, Graz
Semprich, Stephan; Technical University Graz, Graz
Tykal, Jiri; Arsenal Research, Vienna
Wewerka, Manfred; Polyfelt, Linz

Belarus
Sobolevski, Dmitri; Ecotechinvest, Minsk

Belgium
Alboom, Gauthier van; Min. v.d. Vlaamse Gemeenschap, Zwijnaarde
Beknel, Johan; Identity, Mol
Bottiau, Maurice; Socofonda, Brussels
De Cock, Flor; Geo.Be, Lennik
Gille, Henri; Bachy, Bruxelles
Goris, Thierry; Royal Military Academy, Brussels
Haegeman, Wim; Ghent University, Zwijnaarde
Holeyman, Alain; UCL, Louvain-La-Neuve
Imbo, Robert; Franki Geotechnics, Machelen
Impe, Peter van; Ghent University, Zwijnaarde
Impe, William van; Ghent University, Zwijnaarde
Jewell, Richard; Sage Engineering, Brussels
Lecot, Johan ; Dredging International, Zwijndrecht
Legrand, Christian; C.S.T.C/W.T.C.B., Bruxelles
Maertens, Jan; Jan Maertens BVBA, Beerse

List of delegates

Maertens, Luc; Besix, Brussels
Mazzieri, Francesco; Ghent University, Zwijnaarde
Menge, Patrick; Min. v.d. Vlaamse Gemeenschap, Zwijnaarde
Miller, Jean-Pierre; Tractebel Development Eng., Brussels
Van Den Broeck, Marc; Dredging International, Zwijndrecht
Verhoeven, Nadia; I.D. Fos Research E.E.I.G, Geel
Welter, Philippe; Ministere Wallon Equipement, Liege

Bolivia
Salinas, Mauricio; Univ. Mayor De San Simon, Cochabamba

Brazil
Bogossian, Francis; ISSMGE,Rio de Janeiro
Mello, Victor de; Victor F.B. de Mello Ass., Sao Paulo

Canada
Bermingham, Patrick; Berminghammer, Hamilton
O'Brien, Jim; Macleod Geotechnical, North Vancouver, B.C.
Phillips, Ryan; Memorial University, St. John's

Peoples Republic of China
Novais-Ferreira, Henrique; LECM, Macau
Pang, Richard; Govt. of the Hk Spec. Admin. Reg., Hong Kong
Zhu, Guofo; Polytech University, Hong Kong

Colombia
Suarez, Jaime; Univ. Industrial de Santander, Bucaramanga

Croatia
Lisac, Zvonimir; Inst. Gradevinarstva Hrvatske, Zagreb
Maric, Bozica; Conex, Zagreb
Mulabdic, Mensur; Civil Eng. Faculty Osijek, Zagreb
Stanic, Bogdan; Inst. Gradevinarstva Hrvatske, Zagreb
Szavits-Nossan, Antun; University Of Zagreb, Zagreb

Czech Republic
Bohac, Jan; Charles University, Praha
Feda, Jaroslav; Itam Cas, Praha
Koudelka, Petr; Academy of Sc. Czr-Itam, Prague
Lamboj, Ladislav; Cvut-Fac. of Civil Eng., Praha
Skopek, Jiri; Reat, Praha
Vanicek, Ivan; Cvut-Fac. of Civil Eng., Praha

Denmark
Bjerregaard Hansen, Per; Danish Geotechnical Inst., Lyngby
Bodker, Lars; State Port Authoritity, Frederikshavn
Brinch Clausen, Jens; Danish Geotechnical Inst., Lyngby
Denver, Hans; Danish Geotechnical Inst., Lyngby
Gordon, Anne; Comet, Vedbaek
Gormsen, Claus; Niras, Alleroed
Gravgaard, Jens; Ramboll, Virum
Hededal, Ole; Cowi Consulting Eng. & Planner, Lyngby
Jackson, Peter; Comet, Copenhagen
Krebs Ovesen, Niels; Danish Geotechnical Inst., Lyngby
Krijger Hansen, Henning; Danish Geotechnical Inst., Lyngby
Larsen, Hugo; Danish Geotechnical Inst., Lyngby
Olsen, Mogens; Carl Bro As, Glostrup
Sorensen, Carsten; Cowi, Aalborg

Steenfelt, Jorgen; Aalborg University, Aalborg

Egypt
Hamza, Mamdouh; Hamza Associates, Cairo

Estonia
Mets, Mait; Mikrovai, Tallinn

Finland
Arkima, Olli; Ys-Yhdyskunta Oy, Espoo
Eronen, Sami; Rautaruukki, Hameenlinna
Forsman, Juha; Vaitex, Espoo
Gustavsson, Henry; Helsinki Univ. of Techn., Hut
Hakulinen, Matti; Geomatti Oy, Lappeenranta
Hartikianen, Jorma; Tampere University of Techn., Tampere
van den Heuvel, Derk; Junttan, Kuopio
Kolisoja, Pauli; Tampere University of Techn., Tampere
Korkeakoski, Pasi; Tampere University of Techn., Tampere
Lehtonen, Jouko; Rautaruukki, Kaarina
Makela, Harri; Innogeo, Helsinki
Nirhamo, Jarmo; LT-Consultants, Helsinki
Palocahti, Anton; Kaitos Oy, Helsinki
Rathmayer, Hans; VTT Techn. Research Centre, Vtt
Ravaska, Olli; University of Oulu, Oulu
Saarelainen, Seppo; VTT Communities & Infrastruc., Espoo
Salmanhaara, Pekka; De Neef Finland, Helsinki
Sorkamo, Martti; Talentek, Vaasa
Tanska, Harri; Vaitek, Espoo
Tervonen, Timo; Junttan, Kuopio
Tolla, Panu; Finnish National Road Admin., Helsinki
Tuhola, Markku; Utt Comm. & Infrastructure, Utt
Uotinen, Veli-Matti; Finnish National Road Admin., Helsinki
Vahaaho, Ilkka; Helsinki City Real Estate Dep., Helsinki
Valkeisenmaki, Aarno; Finnish National Road Admin., Helsinki
Vepsalainen, Pauli; Helsinki Univ. Of Techn., Espoo
Vuola, Pekka; VR-Track, Helsinki

France
Amar, Samuel; LCPC, Paris
Bardot, Francis; Cabinet d'Expertises Bardot, Lyon
Baudoux, Emmauel; Apageo-Segelm, Magny Les Hameaux
Berthelot, Patrick; Bureau Veritas, Paris
Biarez, Jean; Ecole Centrale Paris, Chatenay Malabry
Blivet, Jean-Claude; Lab. Des Ponts Et Chaussees, Grand Quevilly
Boulon, Marc; University Joseph Fourier, Grenoble
Bretelle, Sylvie; Terrasol, Montreuil
Bustamante, Michel; L.C.P.C., Paris
Cassan, Maurice; Fondasol-Etudes, Avignon
Cui, Yujun; Ecole Des Ponts Et Chaussees, Marne La Vallee
Debats, Jean-Marc; Vibroflotation Sarl, Eguilles
Delage, Pierre; Ecole Des Ponts Et Chaussees, Marne La Vallee
Delattre, Luc; LCPC, Paris
Dore, Michel; Mecasol, Rungis
Durville, Jean-Louis; Lab. Des Ponts Et Chaussees, Paris
Flavigny, Etienne; Universite Joseph Fourier, Grenoble
Fleureau, Jean-Marie; Ecole Centrale Paris, Chatenay-Malabry
Fontaine, Laurent; Terre Armee Internationale, Le Pecq
Frank, Roger; Ecole Des Ponts Et Chaussees, Marne La Vallee

Gambin, Michel; Apageo-Segelm, Paris
Gourves, Roland; Lermes/Cust Clermont. Ferrand, Aubiere
Guerpillon, Yves; Scetauroute, Seyssins
Guilloux, Alain; Terrasol, Montreuil
Kastner, Richard; Insa, Villeurbanne
Leca, Eric; Scetauroute, Pringy
Lefebvre, Francois; Fondasol-Etudes, Avignon
Liausu, Philippe; Menard Soltraitement, Nozay
Magnan, Jean-Pierre; L.C.P.C., Paris
Mermet, Jean-Pascal; Apageo-Segelm, Magny Les Hameaux
Mestat, Philippe; L.C.P.C., Paris
Oberreiter, Klaus; Bidim Geosynthetics, Bezons
Pignerol, Francois; Intrafor, Saint Quentin En Yvelines
Puech, Alain; Fugro France, Nanterre
Quibel, Alain; C.E.T.E. Normandie-Centre, Le Grand Quevilly
Robert, Jacques; Simecsol, Sevres
Roussoulieres, Antonio; Lafarge, St. Quentin Fallavier
Sanglerat, Guy; Lyon
Savasta, Paolo; Setsol, Velaux
Schlosser, Francois; Terrasol, Montreuil
Soubra, Abdul-Hamid; Ensais, Strasbourg
Viggiani, Gioacchino; Universite Joseph Fourier, Grenoble

Germany
Alexiew, Dimiter A.; Huesker Synthetic, Gescher
Arslan, Ulvi; Techn. Univ. Darmstadt, Darmstadt
Berhorst, Volker; Wille Geotechnik, Gottingen
Brunner, Wolfgang; Bauer Spezialtiefbau, Schrobenhausen
Bunjevac, J.M.; Micronic, Berlin
Chambosse, Gerhard; Keller Grundbau, Bochum
Ebersbach, Falk; Bougrund Stralsund Ing.Gesell., Stralsund
Egle, Theo; Bauer Specialtiefbau, Schrobenhausen
Ehrenberg, Henning; Naue Fasertechnik, Espelkamp-Fiestel
El-Mossallamy, Yasser; Arcadis Trischler & Partner, Darmstadt
Erichsen, Claus; Wbi Prof. Dr.-Ing. W. Wittke, Aachen
Evers, Jan; Interfels, Bad Bentheim
Festag, Gerd; Technische Univ. Darmstadt, Darmstadt
Giere, Johannes; Technische Univ. Darmstadt, Darmstadt
Grubert, Peter; Ggu, Braunschweig
Heerten, Georg; Naue Fasertechnik, Luebbecke
Heineke, Stefan; Technische Univ. Darmstadt, Darmstadt
Hestermann, Uwe; Bilfinger & Berger Bau, Mannheim
Horstmann, Johann; Naue Fasertechnik, Luebbecke
Jaup, Achim; University Of Kassel, Kassel
Kassner, Jurgen; Huesker Synthetic, Gescher
Katzenbach, Rolf; Technische Univ. Darmstadt, Darmstadt
Kayser, Jan; Bfw, Hamburg
Klapperich, Herbert; TU Bergakademie Freiberg, Freiberg
Koehler, Hans-Jurgen; Baw Karlsruhe, Karlsruhe
Kruger, Peter; Prof. Peter Kruger Ing.Buro, Ottersberg-Posthausen
Kuehne, Manfred; Dmt, Essen
Ladyjenski, Igor; Huesker Synthetic, Gescher
Mallwitz, Karl; Fh Neubrandenburg, Neubrandenburg
Menzel, Elke; Ggu, Braunschweig
Neff, Hermann; Etn Tropp-Neff & Partner, Hungen/Hessen
Nods, Max; Huesker Synthetic, Gescher
Nussbaumer, Manfred; Ed. Zublin, Stuttgart
Oelckers, Gunter; Oelckers & Steltner, Hamburg

Elfriede, Ott; University Of Kassel, Kassel
Otto, Jorg; Ggu, Braunschweig
Pekoll, Oskar; Leonhardt, Andra & Partner, Berling
Quick, Hubert; Ing.Soz. Katzenbach Und Quick, Darmstadt
Raeker, Hans-Martin; Witt & Jehle Geotechnik, Koblenz
Raithel, Marc; University Of Kassel, Kassel
Reuter, Ernst; Naue Fasertechnik, Luebbecke
Richter, Thomas; Gvd Consult, Berlin
Sadgorski, Willi; Bay.Landesamt Wasserwirt., Munchen
Schmidt-Schleicher, Hermann; Zerna, Kopper & Partner, Bochum
Schuppener, Bernd; Baw, Karlsruhe
Sobolewski, Janusz; Huesker Synthetic, Gescher
Sprick, Gabriele; Naue Fasertechnik, Lubbecke
Stocker, Manfred; Bauer Spezialtiefbau, Schrobenhausen
Tirpitz, Ernst-Rainer; Bilfinger & Berger Bau, Mannheim
Trolle, Christine; Keller Grundbau, Offenbach
Turek, Jens; Technische Univ. Darmstadt, Darmstadt
Vermeer, Pieter; Universitat Stuttgart, Stuttgart
Wieners, Andreas; Hsp Hoesch Spundwand U. Profil, Dortmund
Wille, Thorsten; Wille Geotechnik, Gottingen
Wittke, Walter; Wbi Prof.Dr.-Ing. W. Wittke, Aachen
Zaeske, Dirk; University Of Kassel, Kassel
Ziegler, Ralf; Naue Fasertechnik, Adorf/Vogtl.

Greece
Anagnostopoulos, Andreas; Nat. Techn. Univ. of Athens, Kifissia
Anastassopoulos, Konstantin; Public Power Corp. Greece, Athens
Cavounidis, Spyros; Edafos, Athens
Dounias, George; Edafos, Athens
Fikiris, Ioannis; Edafos, Athens
Gazelas, Dimitrios; Sgi Hellas, Athens
Houssiadas, Vassilis; Egnatia Odos, Thermi
Karkanias, Sotirios; Athens
Kavvadas, Michael; Nat. Techn. Univ. of Athens, Kifissia
Loizos, Andreas; N.T.U.A., Athens
Mihalis, Ilias; Nat. Techn. Univ. of Athens, Athens
Saffari, Nader; Sir Alexander Gibb, Athens
Vlavianos, George; Ministry of Public Works, Papagou

Hong Kong
NG, Charles; Univ. Of Sciences & Techn., Hongkong

Hungary
Imre, Emoke; TU of Budapest, Budapest
Mecsi, Jozsef; Me-Szi Engineering, Budapest
Szepeshazi, Robert; Szechenyi Istvan College, Gyar

Iceland
Skulason, Jon; Almenna Verkfraedistofan, Reykjavik

Ireland
Creed, Michael; University College Cork, Cork
Creighton, Ronnie; Geological Survey of Ireland, Dublin
Farrell, Eric; Trinity College, Dublin
Lehane, Barry; Trinity College, Dublin
Luby, Derek; John Barnett And Associates, Dublin
Orr, Trevor; Trinity College, Dublin

Israel
Zolkov, Eli; Ramat Hasharon

Italy
Barla, Giovanni; Politecnico di Torino, Torino
Belviso, Renato; Politecnico di Bari, Bari
Briganti, Renato; Italferr, Roma
Buonanno, Alessandro; Ferrovie dello Stato, Roma
Calabresi, Giovanni; Universita di Roma La Sapienza, Roma
Contini, Massimo; Pagani Geotechnical, Calendasco
Federico, Antonio; Politechnico, Taranto
Fiocchi, Viviana; Sisgeo, Segrate (Mi)
Fratalocchi, Evelina; University of Ancona, Ancona
Goretti, Massimo; Soil Test, Arezzo
Jamiolkowski, Mike; Techn. University of Torino, Milano
Marchetti, Silvano; L'Aquila University, Roma
Marchisella, Raffaele; Italferr, Roma
Marco, Martino di; Controls Srl, Cernusco Sul Naviglio Mi
Maugeri, Michele; University Of Catania, Catania
Monaco, Paola; L'Aquila University, L'Aquila
Napoleoni, Quintius; University of Rome La Sapienza, Rome
Pasqualini, Erio; University of Ancona, Ancona
Quieti, Mauro; Sisgeo, Segrate (Mi)
Russo, Gianpiero; Universita Federico Ii Napoli, Napoli
Scarpelli, Giuseppe; University of Ancona, Ancona
Soranzo, Maurizio; Universita di Udine, Udine
Stella, Massimo; University of Ancona, Ancona
Tiziano, Casagrande; Ferrovie Dello Stato, Rome
Valla, Luigi; Pagani Geotechnical, Calendasco
Vittorio, Misano; Ferrovie Dello Stato, Rome

Japan
Akai, Koichi; Osaka Soil Test Lab., Kyoto
Hayashi, Yoshihiro; Raito Kogyo, Tokyo
Ishihara, Kenji; Science University Of Tokyo, Tokyo
Isobe, Kaneharu; Raito Kogyo, Tokyo
Matsuda, Hiroshi; Yamaguchi University, Yamaguchi
Matsui, Kenji; Cti Engineering, Fukuoka
Matsumoto, Tatsunoti; Kanazawa University, Kanazawa
Mimura, Mamoru; Kyoto University, Gokasho Uji
Nakano, Ryoki; Gifu University, Gifu-Shi, Gifu-Pref.
Ochiai, Hidetoshi; Kyushu University, Fukuoka
Oka, Fusao; Kyoto University, Kyoto
Otani, Jun; Kumamoto University, Kumamoto
Saitoh, Kunio; Nikken Sekkei, Kawasaki
Sekiguchi, Hideo; Kyoto University, Kyoto
Shibata, Azuma; Kowa Corp., Niigata City
Shogaki, Takaharu; Nat. Defense Academy, Yokosuka, Kanagawa
Tanaka, Yasuo; Kobe University, Kobe
Tsuzuki, Makoto; Fugro Japan, Tokyo

Kazakhstan
Idrisov, Dinmuhammed; Alma-Cil, Almaty
Teltayev, Bagdat; Nat. Academy of Trans. & Comm., Almaty
Zhusupbekov, Askar; Karaganda Metallurgical Inst., Temirtau

Korea, Republic
Kim, Sang-Kyu; Dong-Pusan College, Pusan

Lee, Myung Whan; Piletech Consulting Engineers, Seoul

Latvia
Celmins, Valters; Cbkpb, Riga

Lithuania
Furmonavicius, Liudvikas; Vilnius Techn. University, Vilnius
Gadeikis, Saulius; Vilnius University, Vilnius
Kudzys, Antanas; Inst. of Architect. & Constr., Kaunas
Slauteris, Arturas; Geoprojektas, Stc, Klaipeda
Stragys, Vincentas; Vilnius Gediminas Techn. Univ., Vilnius

Luxembourg
Casasanta, Pietro; Profilarbed-Ispc, Esch-Sur-Elzette
Jacoby, Edmond; Profilarbed-Ispc, Esch-Sur-Alzette
Schmitt, Alex; Ispc, Esch-Sur-Alzette

Macedonia, Rep. Of
Dimitrievski, Wupco; Geing, Skopje

Malaysia
Lawson, Chris; Royal Ten Cate, Pesaling Jaya

Mexico
Springall, Guillermo; Geotec, Mexico

Morocco
Ben Azzouz, Nabil; Lab. Public d'Essais & d'Etudes, Casablanca
Ejjaaouani, Houssing; L.P.E.E., Casablanca

Netherlands
Adel, Jeroen van den; Universiteit Twente, Enschede
Asten, Pieter van; A.P. van den Berg, Heerenveen
Baars, Stefan van; Strukton T+E Consult, Nieuwegein
Bakker, Klaas; Bouwdienst Rijkswaterstaat, Utrecht
Barends, Frans; Geodelft, Delft
Barneveld, Albert; Dienst Weg- en Waterbouwkunde, Delft
Beernink, E.; Diana Analysis, Delft
Beetstra, Gerben; Geodelft, Delft
Bellis, Jon; NBM Amstelland/Dirk Verstoep, Rotterdam
Berg, Martin van den; Terre Armee-Reinforced Earth, Breda
Berg, Peter van den; Geodelft, Delft
Bezuijen, Adam; Geodelft, Delft
Bierman, Willem; TU Delft, Delft
Bleumer, Sonja; Geodelft, Delft
Blommaart, Peter; Ministry of Public Works, Delft
Boer, Freerk de; Holland Railconsult, Utrecht
Bom, Sacha; Plaxis, Delft
Bout, L. van den; Diana Analysis, Delft
Brand, Peter; Plaxis, Delft
Bremmer, Chris; TNO, Delft
Brinkgreve, Ronald; Plaxis, Delft
Brugman, Marijn; Fugro Engineers, Leidschendam
Burger, Peter; Fugro Ingenieursbureau, Leidschendam
Chen, Li Jen; IHE, Delft
Collignon, Tom; Mos Grondmechanica, Rhoon
Cornejo, Carlos; Delft Univ. of Technology, Delft
Cortlever, Nico; Geotechnics, Amsterdam

List of delegates

Daalen, Paul van; Aveco, Utrecht
Dalen, Jan van; Gemeentewerken Rotterdam, Rotterdam
De Jong, Baukje; Ham, Rotterdam
Deen, Jurjen van; Geodelft, Delft
Docters Van Leeuwen, Linda; Aveco, Utrecht
Duijvenbode, Jan Dirk van; Rijkswaterstaat, Delft
Duskov, Milan; Oranjewoud Infragroep, Capelle a/d IJssel
Duyfjes, Arne; A.P. van den Berg, Heerenveen
Eijgenraam, Sander; Holland Railconsult, Utrecht
Elzen, Martijn van den; T & E Consult, Maarssen
Esposito, Genarro; TNO Bouw, Delft
Feijter, Jan de; Geodelft, Delft
Felix Eceberre, Jose Maria; IHE, Delft
Foeken, Rob van; TNO Bouw, Delft
Goudoever, Piet van; Fugro Ingenieursbureau, Leidschendam
Graaf, Henk van de; Sonar Geotechnical Eng., Son
Grashuis, Arjan; Public Works & Water Management, Delft
Groot, Claes; Holland Railconsult, Utrecht
Groot, Maarten de; Geodelft, Delft
Geerhard, Hannink; Gemeentewerken Rotterdam, Rotterdam
Heer, Ronald de; IHE, Delft
Heijmans, P.J.M.J.; Terre Armee, Breda
Heijnen, Wim; Roosendaal
Helbo, Tim; Van Oord Acz, Gorinchem
Herbschleb, Jurgen; De Weger Arch. & Cons. Eng., Rotterdam
Ho, Hsin-Yenia; IHE, Delft
Hoang Vi, Minh; Delft University of Technology, Delft
Hoefsloot, Flip; Fugro Ingenieursbureau, Leidschendam
Hogenes, Kees; Omegam, Amsterdam
Hooydonk, W. van; Geomil Equipment, Alphen a/d Rijn
Horst, Aad van der; Holl. Beton en Waterbouw, Gouda
Huiden, Evert; NGT, Gouda
Hutteman, Marco; Mos Grondmechanica, Rhoon
Jaarsveld, Erik van; Geodelft, Delft
Jansen, Hein; Fugro Ingenieursbureau, Leidschendam
Jong, Rolf de; Mos Grondmechanica, Rhoon
Jonker, Fred; CUR, Gouda
Joustra, Kees; Leidschendam
Kay, Steve; Fugro Engineers, Leidschendam
Kenkhuis, Jan; Geotechnical Support, Almere
Kimura, M.; Geodelft, Delft
Kleinjan, Arnold; Gemeentewerken Rotterdam, Rotterdam
Knibbeler, Alexander; Mos Grondmechanica, Rhoon
Koehorst, Benno; Rijkswaterstaat, Delft
Koelewijn, Andre; Delft Univ. Of Technology, Delft
Koers, Carla; TNO, Delft
Kolk, Harry; NGT, Gouda
Kolk, Harry, Fugro Engineers, Leidschendam
Kooistra, Annemarije; Ballast Nedam, Amstelveen
Kooperen, Kees van; Badhoevedorp
Kort, Arjen; Delft Univ. of Technology, Delft
Kosgoda, K.M.S.K.; IHE, Delft
Kruizinga, Jan; Koac. Wmd, Apeldoorn
Kuile, S. ter; Geodelft, Delft
Langhorst, Peter; NGT, Gouda
Leendertse, Wim; Proj. Org. High Speed Line South, Utrecht
Leeuw, Bert de; Delft
Leijen, Walter; IHC Hydrohammer/Fundex, Kinderdijk

Lierop, Sylvia van; Strukton Betonbouw, Maarssen
Lindenberg, Jaap; Ministry of Transport, Delft
Lu, Chih Wei; IHE, Delft
Luger, Dirk; Geodelft, Delft
Maurenbrecher, ;Michiel, Delft Univ. of Technology, Delft
Meer, Martin v.d.; Fugro Ingenieursbureau, Leidschendam
Mlynarek, Zbigniew; A.P. van den Berg, Heerenveen
Molendijk, Waldo; Geodelft, Delft
Nagtegaal, Joop; Strukton Betonbouw, Maarssen
Nekeman, Sebastian; Delft Univ. of Technology, Delft
Nelissen, Henk; Geodelft,Delft
Nieuwenhuis, Jan; TU Delft, Delft
Nohl, Wim; Fugro Ingenieursbureau, Leidschendam
Nutbroek, Jos; Boart Longyear, Etten-Leur
Obladen, Bas; Ballast Nedam, Beton & Water, Amstelveen
Omtzigt, E., Geomil Equipment, Alphen a/d Rijn
Oosterhout, Geert van; TNO, Delft
Oostveen, Jack P.; Delft Univ. of Technology, Delft
Os, Patrick van; Ballast Nedam, Amstelveen
Otter, Remmert den, Technosoft, Lochem
Pachen, Harry; Public Works Rotterdam, Rotterdam
Paesschen, Jack; Arbed Damwand, Moerdijk
Paff, Gert-Jan; Fugro Engineers, Leidschendam
Peuchen, Joek; Fugro Engineers, Leidschendam
Pongpothakul, Chana; IHE, Delft
Pontier, Adrian; Holland Railconsult, Utrecht
Poulson, Stephen; Rotterdam
Quelerij, Louis de; Fugro Ingenieursbureau, Leidschendam
Revoort, E.; Fund. Technieken Verstraeten, Oostburg
Roos, Leo; Fugro Engineers, Leidschendam
Saathof, Koos; Ministry of Transport, Delft
Scholey, Graham; Fugro Engineers, Leidschendam
Schotmeyer, Gert-Jan; Geodelft, Delft
Schrier, Joost van; Haskoning, Nijmegen
Seters, Adriaan van; Fugro Ingenieursbureau, Leidschendam
Shinn, J.; Geomil Equipment, Alphen a/d Rijn
Six, Bert; Inpijn-Blokpoel, Son
Sluimer, Eric; Ingenieursbureau Amsterdam, Amsterdam
Smits, Maarten; Fugro Ingenieursbureau, Leidschendam
Sri Susiswo, Herlambang; IHE, Delft
Stevelink, Walter; Reeuwijk,
Strack, Erik; Delft Univ. of Technology, Delft
Stuit, Herke; Holland Railconsult, Utrecht
Suijs, P.W.; Spanbeton, Koudekerk a/d Rijn
Suiker, R.; Geodelft, Delft
Termaat, Ruud; Ministry of Transport, Delft
Teunissen, Egbert; Witteveen & Bos Consulting Eng, Deventer
Theuns, Cees; Naue Benelux, Dongen
Tigchelaar, Jan; Geodelft, Delft
Tjaden, J.H.; Adviesbureau Tjaden, Haarlem
Tol, Frits van; Gemeentewerken Rotterdam, Rotterdam
Torstensson, Bengt; Geotechnics, Amsterdam
Tsai, Tsung Chuan; IHE, Delft
Tschuschke, Wojciech; A.P. van den Berg, Heerenveen
Twillert, Jan Aart van; Geodelft, Delft
Uelman, Frans, Hydronamic, Papendrecht
Van, Meindert; Geodelft, Delft
Veghel, Maurice van; Fugro Ingenieursbureau, Leidschendam

Velde, Eelco van der; Dirk Verstoep Fund. Techniek, Gorinchem
Venmans, Arjan; Min. of Transp. & Public Works, Delft
Verruijt, Arnold; Delft Univ. of Technology, Delft
Visser, Gerard; Omegam, Amsterdam
Vriend, Ad; Ballast Nedam Fund. Technieken, Dordrecht
Vrolijk, A.; Nedeximpo, Amsterdam
Wang, Ben Jin; IHE, Delft
Wassenaar, E.; Geomil Equipment, Alphen a/d Rijn
Weele, Bram van; IFCO-Funderingsexpertise, Stolwijk
Weele, Frans van; IFCO, Waddinxveen
Weesep, Ben van; Fundamentum, Sliedrecht
Welling, Engbert; A.P. van den Berg, Heerenveen
Wentink, Joost; Geodelft, Delft
Wiel, J.J. van der; Fund. Technieken Verstraeten, Oostburg
Wit, Theo de; Joustra Geomet, Alphen a/d Rijn
Woldringh, Bob; DHV Environm. & Infrastructure, Amersfoort
Zigterman, Wolter; ITC-Delft, Delft
Zuidberg, Herman; Fugro Engineers, Leidschendam
Zwaag, Gerrit van der; Fugro Engineers, Leidschendam

Nigeria
Akwuba, Godwin; Progress Engineers, Lagos
Folayan, Joseph; Progress Engineers, Lagos
George, Enoch; E. George Associates, Port Harcourt

Northern Ireland
Mccandless, Gary; Construction Service, Belfast

Norway
Andersen, Knut; Norwegian Geotechnical Inst., Oslo
Haarvik, Linda; Norwegian Geotechnical Inst., Oslo
Hermann, Steinar; Norwegian Geotechnical Inst., Oslo
Karlsrud, Kjell; Norwegian Geotechnical Inst., Oslo
Kulsrud, Harald; Geonor, Oslo
Kvalstad, Tore; Norwegian Geotechnical Inst., Oslo
Lacasse, Suzanne; Norwegian Geotechnical Inst., Oslo
Leirvik, Karl; Geonor, Oslo
Lunne, Tom; Norwegian Geotechnical Inst., Oslo
Madshus, Christian; Norwegian Geotechnical Inst., Oslo
Simonsen, Arne; Noteby, Oslo

Poland
Bednarek, Roman; Technical University, Szozecin
Dembicki, Eugeniusz; Technical University Gdansk, Gdansk
Kazimierczak, Jan; Energopol, Szczecin
Lechowicz, Zbigniew; Warsaw Agricultural University, Warsaw
Rymsza, Bogdan; Warsaw Univ. of Technology, Warszawa
Tejchman, Andrzej; Technical University Gdansk, Gdansk
Topolnicki, Michal; Techn. Univ. of Gdansk, Gdansk
Ukleja, Kazimierz; Poltegor Institute, Wroclaw
Wolski, Wojciech; Warsaw Agricultural Univ., Warszawa

Portugal
Almeida Mendes, Joaquime; Brisa Auto-Estradas Portugal, Sao Domingos De Rana
Correia, Rui Manuel; LNEC, Lisboa
Cruz, Jose; Geosolve, Lisbon
Ferreira Martins, Francisco; Universidade Di Minho, Guimaraes
Fortunato, Eduardo; Laboratorio National Eng.Civil, Lisboa

Gomes-Correia, Antonio; Techn. Univ. Of Lisbon/Ist, Lisboa
Guerra, Nuno; Techn. Univ. Of Lisbon, Lisboa
Lopes, M-Lurdes; Feup, Porto
Lourenco Cardozo, Mario Antonio; Brisa Auto-Estradas Portugal, Sao Domingos De Rana
Maranha Das Neves, Emanuel; Estado Das Obras Publicas, Lisboa
Marques Dos Santos, Jorge Santos; Tecnasol, Amadora
Mateus De Brito, Jose Antonio; Cenor, Lisboa
Oliveira Domingues, Ernesto Manuel; Brisa Auto-Estradas Portugal, Sao Domingos De Rana
Pinto, Alexandre; Tecnasol, Amadora
Seco E Pinto, Pedro; Laboratorio National Eng. Civil, Lisboa

Romania
Chirica, Anton; Techn. Univ. Bucharest, Bucharest
Manea, Sandra; Techn. Univ. for Civ. Eng, Bucharest
Manoliu, Iacint; Techn. Univ. of Civil Eng., Bucharest
Marcu, Anatolie; Techn. Univ. of Civil Eng., Bucharest
Nabosnyi, Atila; Conel-Hidroelectrica, Bucuresti
Popescu, Mihail; Univ. of Civil Engineering, Bucharest
Radulescu, Nicoleta; T.U.C.E. Bucharest, Bucharest
Raileanu, Paulica; Technic University, Iasi

Russia
Aleynikov, Sergey; St.Academy of Arch.& Construc., Voronezh
Bellendir, Evgeny; B.E. Vedeneev Vniig, St. Petersburg
Ilyichev, Vyacheslav; Niiosp, Moscow

Saudi Arabia
Aiban, Saad; Kfupm, Dhahran

Singapore
Molnit, Thomas; Loadtest International, Singapore
Toll, David; Nanyang Techn. University, Nanyang

Slovakia
Slavik, Ivan; Slovak Univ. of Technology, Bratislava
Turcek, Peter; Slovak Univ. of Technology, Bratislava

Slovenia
Brozic, Dusanka; Geot, Ljubljana
Demsar, Vladimir; Building & Civil Eng. Inst., Ljubljana
Gaberc, Ana; Fac.Of Civil & Geodetic Eng., Ljubljana
Kovacic, Andreja; Gracen D.O.O., Ljubljana
Pulko, Bostjan; Fac. of Civil & Geodetic Eng., Ljubljana
Zeleznik, Drago; Sct D.D. Cp, Ljubljana
Zvanut, Pavel; Nat. Building & Civil Eng. Inst., Ljubljana

South Africa
Day, Peter; Jones & Wagener, Rivonia

Spain
Alonso, Eduard; Dit. Upc, Barcelona
Burbano, German; Dragados, Madrid
Cuellar, Vicente; Cedex, Madrid
Dapena, Enrique; Lab. Geotechnia Cedex, Madrid
Del Canizo Perate, Zuis; Esteyco, Madrid
Fort, Luis; Necso, Aledbendas
Gens, Antonio; Techn. University of Catalunya, Barcelona
Lloret, Antonio; Techn. Univ. of Catalonia, Barcelona

Nuche, Ignacio; Ineco, Madrid
Rodriguez Ballesteros, Fernando; Fomento De Constr. Y Contratas, Madrid
Sagaseta, Cesar; Universidad De Cantabria, Santander

Sweden
Alen, Claes; NCC, Goteborg
Andreasson, Bo; J & W, Goteborg
Bengtsson, Per-Evert; Swedish Geotechnical Inst., Linkoping
Berggren, Bo; SGI, Linkoping
Caprona, Guy de; Geotech, Askim
Ekenberg, Mats; Skanska Teknik, Gothenborg
Hartlen, Jan; Lund University, Lund
Holm, Goran; Swedish Geotechnical Inst., Linkoping
Jendeby, Leif; NCC, Goteborg
Johansson, Bo; Chalmers Univ. of Techn., Goteborg
Larsson, Stefan; Tyrens Infrakonsult, Stockholm
Lundahl, Bjorn; SCC Project Management, Stockholm
Massarsch, Rainer; Geo Engineering, Bromma
Middendorp, Peter; TNO Profound, Rijswijk
Oberg-Hogsta, Anna-Lena; Chalmers Univ. of Techn., Goteborg
Olsson, Connie; SGI, Linkoping
Phung, Doc Long; Stabilator, Stockholm
Rydell, Bengt; Swedish Geotechnical Inst., Linkoping
Sundquist, Owe; Borros Ab, Solna
Svensson, Lennart; J & W Ab Jacobson & Widmark, Goteborg
Viberg, Leit; Swedish Geotechnical Inst., Linkoping

Switzerland
Amann, Peter; Eth-Zurich (Honggerberg), Zurich
Brenner, Peter; Weinfelden
Dysli, Michel; Ecole Polytechnique Federale, Lausanne
Egger, Peter; Ecole Polyt. Fed. de Lausanne, Lausanne
Naterop, Daniel; Solexperts, Schwerzenbach
Schneider, Hans; Geo-Consulting, Zug
Thut, Arno; Solexperts, Schwerzenbach
Vulliet, Laurent; Epfl, Lausanne

Thailand
Paochaiyangyen, Rutian; Fugro-Ign (Thailand), Klongtoey, Bangkok
Ruenkrairergsa, Teeracharti; Road Research & Dev. Centre, Bangkok

Tunisia
Sassi, Abdelhamid; Securas, Tunis

Turkey
Akselik, Necise; Gen. Directorate of Highways, Ankara
Aksoy, Salih; Gen. Directorate of Highways, Ankara
Caglayan, Hidir; Mng Zemtas, Ankara
Catakli, Mahmut; Ozbir Muhendislik, Kisiku/Istanbul
Cinicioglu, S. Feyza; Istanbul University, Istanbul
Croissant, Ricardo; Holzmann-Gama-Strabag C.W.J.V., Nizip (Gaziantep)
Ergun, M. Ufuk; Middle East Technical Univ., Ankara
Erol, Orhan; Middle East Technical Univ., Ankara
Etkesen, Zuhal; Gen. Directorate Of Highways, Ankara
Gok, Sebahat; Istanbul Technical University, Istanbul
Guler, Erol; Bogazici University, Istanbul
Horoz, Attila; Yurcel Proje, Ankara
Kin, Ali Suha; Stfa Temel Pile Cons., Istanbul

Ozben, Ali; Gen. Directorate of Highways, Etller
Ozdemir, Suleyman; Gen. Directorate of Highways, Etller
Tan, Oguz; Techn. Univ. of Istanbul, Istanbul
Togrol, Ergun; Istanbul Technical University, Istanbul
Unal, Gulgun; Anadolu University, Eskisehir

Ukraine
Doubrovsky, Michael; Odessa State Maritime Univ., Odessa

United Arab Emir.
Kamal Mohammed, Fathy; MISR Raymond Foundation, Abu Dhabi

United Kingdom
Anderson, Bill; University of Sheffield, Sheffield
Babin, Pascal; RLE, London
Bloodworth, Alan; University of Oxford, Oxford
Bond, Andrew; Geotechnical Consulting Group, Banstead
Cooper, Michael; Birmingham University, West Midlands
Cooper, Sam; Emap Construct., London
Crilly, Mike; BRE, Watford
Edmonds, Helen; Geotechnical Consulting Group, London
Gourvenec, Susan; Cambridge University, Cambridge
Greenwood, David; Gerrards Cross
Grose, William; Ove Arup & Partners, London
Heelis, Michael; Nottingham University, Nottingham
Herr, Monika; John Wiley, Chichester
Hiller, David; Transport Research Laboratory, Berks
Houlsby, Guy; Oxford University, Oxford
Hughes, John; Tensar International, Blackburn
Jardine, Richard; Imperial College, London
Jas, Hein; Tensar International, Blackburn
Landtsheer, Jan de; John Wiley, Chichester
Linney, Lionel; Montgomery Watson, Bucks
Lord, Andrew; Ove Arup & Partners, London
Mair, Robert; Cambridge University, Cambridge
Menzies, Bruce; GDS Instruments, Egham, Surrey
Myles, Bernard; Groupe T.A.I., Hemel-Hempstead
Parry, Dick; ISSMGE/Univ. Eng. Dept., Cambridge
Plant, Graham; Berkhemsted, Herts
Powell, John; BRE, Watford
Pyrah, Ian; Napier University, Edinburgh
Rankin, William; Mott Mac Donald, Croydon
Rogers, Chris; University of Birmingham, Birmingham
Rudrum, Mark; Ove Arup & Partners, London
Rulens, Dominique; Rail Link Engineering, London
Simpson, Brian; Ove Arup & Partners, London
Soudain, Max; Emap Construct., London
Sutton, Jerry; Gds Instruments, Surrey
Taylor, Neil; City University, London
Thurlwell, Paul; Tensar International, Blackburn
Tonks, David; Edge Consultants UK, Manchester
Vaughan, Peter; Ipswich,
Vijting, Bon; Tensar International Blackburn
Waterman, Keith; Kvaerner Cementation Found., Rickmansworth
Wheeler, Paul; Emap Construct., London

United States Of America
Chacko, Jacob; Fugro, Ventura, Ca

List of delegates

Crawley, Julian, Stent, Vellejo, Ca
Degen, Wilhelm; Layne Christensen, Irvine, Ca
Eliahu, Shalom; Se Consulting, Lafayette, Ca
Hayes, Jack; Loadtest, Gainesville, Fl
Mcneilan, Tom; Fugro, Ventura, Ca
Mesri, Reza; University of Illinios, Urbana, Il
Smith, Ron; Geo-Institute of the ASCE, Las Vegas, Nv
Wahls, Harvey; N.C. State University, Raleigh, Nc
Znidarcic, Dobroslav; University Of Colorado, Boulder, Co